Bob Freitag

Resilient Shores: Adapting to Climate Change Through Storytelling

Springer

Bob Freitag
Department of Urban Design and Planning
College of Built Environments
University of Washington
Seattle, WA, USA

With Contributions by
JLS Clark
Portland, OR, USA

Gerald Murphy
University of Florida
Gainesville, FL, USA

Ron Kasprisin
Department of Urban Design and Planning
College of Built Environments
University of Washington
Seattle, WA, USA

Dave Carlton PE
DKCarlton & Associates
Seattle, WA, USA

ISBN 978-3-032-09046-1 ISBN 978-3-032-09047-8 (eBook)
https://doi.org/10.1007/978-3-032-09047-8

Cover Illustration: Ron Kasprisin. Sketch depicting students preparing for a public workshop held in 1998. 2024.

This Springer imprint is published by the registered company Springer Nature Switzerland AG
The registered company address is: Gewerbestrasse 11, 6330 Cham, Switzerland

When the old Greek Gods ruled, Homer told of a story where the God Apollo enabled Cassandra to foresee perils threatening her City of Troy home. She once tried to warn her townsfolk but was not believed.

In this story, Cassandra lives in Troy Beach. Apollo is a hurricane and the peril it represents is the result of a changing climate. As with the old story, Cassandra warns the town. She is ignored.

Continuing with our metaphor, remember the Trojan horse? Cassandra warned that it would be about the destruction of Troy. She was ignored.

Will Troy Beach collapse, or survive the attack?

Dedicated to Stevie, Atlas, Casper, Emmitt, Isabelle, Lucy, Madeleine, Margot, Mathison, Oscar, Sophie, and all of your grandchildren.

Introduction: A Guided Quest

Artist: Ron Kasprisin

Resilient Shores is a success story. It introduces Troy Beach—a climate-threatened community being torn apart by stronger and increasingly frequent storms.

Are you having difficulty discussing climate change and its potential impacts on you, your family, and your community?

Troy Beach is being swallowed up by rising seas. Yet this community is confronting its risks through storytelling. Conflicts leading to resolutions are presented initially as stories, an approach we have been using our entire lives.

Their quest is personal, emotional, and often visceral. It is not academic, not abstract. This may also be your situation. Creating and telling stories provides a lifeline.

Resilient Shores introduces you to characters not unlike yourselves, who are concerned about their threatened community, their threatened lives. Their story presents context to help you reflect on and debate issues, face conflicts, and suggest resolutions. It doesn't tell, it shows, how extreme and rapid change can be confronted head-on, in a manner that is natural and familiar.

As Porter Abbott introduced in the *Cambridge Introduction to Narrative*, "We make narratives [stories] many times a day, every day of our lives. And we start doing so almost from the moment we begin putting words together. As soon as we follow a subject with a verb, there is a good chance we are engaging in a narrative discourse. "I fell down," the child cries, and in the process tells her mother a little narrative.[1] "We are well-practiced in using stories to resolve foreseen conflicts, identify curative goals, or design pathways. 'I will give my brother a toy and he won't cry, we can have ice cream, and if/when I clean my room I can watch TV.'"

Resilient Shores is a story of how individuals and communities of individuals resolve conflict, and in the process show us that success is possible. Resolutions are shown, and the process should be as familiar as the stories you have been creating and sharing your entire life.

You may feel that similar risks threatening your communities may be nearly impossible to discuss, let alone achieve.

Resilient stories can be:

- Emotionally secure and safe expressions of recovery, reclaiming values threatened by or experiencing life-altering changes.
- Pathways to recovery of a past well-being, something better, or different.

Resilient Shores introduces you to the coastal community of Troy Beach and its impasse as a story. We hope using this process will help you be better prepared to discuss and face climate change directly with all its implications. Whether your community consists of family, friends, or work colleagues, is anchored to a specific physical place or not, many of you will have little choice but to adapt—reducing risks, increasing your ability to bounce back, or reorganizing to a new normal.

Addressing our changing climate can be overwhelming, emotionally draining, and often prohibitively threatening. However, tackling climate change as a **story** may provide the protective therapeutic abstractness necessary to devise difficult achievable resolutions. Feel free to read this as a mystery story. For action, we invite you to create your own story of your future and have included guidance at the end of each chapter. We encourage you to share your stories with your community, revise your characters, settings, problems, and resolutions, and retell them. Whether you create your own stories or read *Resilient Shores* in isolation, remember that storytelling isn't just about words; it is about connecting hearts and minds. Stories have the power to address what might otherwise be too challenging to discuss directly.[2] Storytelling helps to externalize disturbing conflicts and resolve the controversies

[1] Abbott, H. Porter. The Cambridge Introduction to Narrative. Cambridge University Press, 2002.
[2] World Health Organization. The power of storytelling for health impact, 2024.

objectively with a degree of separation.[3] Storytelling can lead to accepting situations beyond one's control.[4,5]

Storytelling can not only be therapeutic, but the process may also direct future behavior. The concept of "Future Self-continuity"[6] suggests that we are more likely to pursue future goals and aspirations if they are included in our personal stories of the future.

Troy Beach is trying its utmost to bounce back from a brutal hurricane. Most permanent and seasonal residents of this coastal community want to bounce back to the normal they had enjoyed—the exception being Cassandra, Troy Beach's newly appointed Planning Director, who has had enough. She was hired to perform the bureaucratic tasks required of the position, not inspiration for a different future. But she has had enough of this bouncing back, only to bounce back again and again as if her community was practicing some never-ending game. But her prophecies are not believed.

Troy Beach isn't any specific community; it is every community threatened by climate change. Like thousands of other communities, Troy Beach is forced to confront extreme heat, rising seas, flooding, drought, and the disappearance of ocean resources.

Are there lessons from experiencing Troy Beach's quest?

Resilient stories can:

- Be personal musings, satisfying in their own right, or
- Spur action.

According to the National Ocean and Atmospheric Administration (NOAA), "By 2050, up to $106 billion worth of coastal property will likely be below sea level if current trends continue. U.S. coasts span the Atlantic, Pacific, and Arctic oceans, the Gulf of Mexico, and the Caribbean Sea. More than 128 million people live in coastal counties, including the Great Lakes coast. Coastal ecosystems support many industries, provide recreational opportunities, and protect communities from storms. Coasts are also important to global trade and in providing job opportunities."[7]

These expected changes will cause a host of problems, but they will also bring opportunities. The rate of climate-influenced change will continue until it reaches a

---

[3] White and Epston. Narrative Means to Therapeutic Ends. W. W. Norton and Company, 2023.

[4] Hayes Steven, Pistorello Jacqueline, and Levin Michael. Acceptance and Commitment Therapy as a Unified Model of Behavior Change. The Counseling Psychologist, 40(7) 976–1002, 2012.

[5] Gaudiano, Brandon. Evaluating Acceptance and Commitment Therapy: An Analysis of a Recent Critique. International Journal of Behavioral Consultation and Therapy, 2011.

[6] Hershfield and Bartels. The Future Self. In Oettingen, G., Sevincer, A.T., & Gollwitzer, P.M. (eds). The Psychology of Thinking about the Future. The Guilford Press, 89–109, 2018.

[7] Climate Change Impacts on Coasts. US EPA, February 2025.

new balance. This is not good for us or the living things we have come to depend on. We must do all we can to reverse this trend.

A changing climate creates the need for more adaptable cities, sustainable farming, places of refuge, green energy, and more. This need will also bring opportunities for new jobs, rewarding experiential quests, and the prospects of creating new adaptable surroundings.

Resilient Shores can be thought of as a guided journal, with a twist. The twist is that journals typically chronicle conditions as they occur. This guided journal is designed to help you create a more resilient future vision for you and your community. It is written to put a face to your generation's most challenging problem. It is more a guided quest than a journal.

Creating and telling stories can make you feel better. They can actually lower your blood pressure,[8] and they can also help resolve conflict by externalizing the problem. Stories can frame conflicts as separate from personal identity; we gain perspective. Sharing personal stories fosters empathy. When we understand others' perspectives, conflict resolution becomes more effective. Storytelling can find common ground. Narratives reveal shared experiences and values, bridging gaps between conflicting parties. They can help shape our choices and interactions.

You can read *Resilient Shores* as a mystery, internalizing presented conflict and resolve, as your expression of your personal story of an externalized reflection and resolution, or as an adventure to share with friends. Who knows? Real achievable solutions might be hiding between the lines. As you proceed through our story of Troy Beach, you are invited to actively participate and explore ways your communities may have to respond.

We have included a screenplay in the appendix to offer variety in your search for solutions. You may read the story and screenplay alone or together with friends and share your mutual experiences as you discover alternative and acceptable pathways forward.

Regardless of how you read *Resilient Shores*, we hope you will engage others in confronting this impenetrable subject—climate-related consequences impacting your lives and those of your family, friends, and communities.

We encourage you to research available adaptation material under such headings as community relocation, and evacuation,[9] managed retreat,[10] hazard[11] and climate mitigation,[12] community adaptation, and storytelling.[13] You may discover valuable objective content, but this is beyond our scope. Our fictional account is not intended

[8] Oettingen, Gabriele. Positive Thinking Inside the New Science of Motivation. Penguin Random House, 2025.

[9] National Academy of Sciences. Community-Driven Relocation: Recommendations for the U.S. Gulf Coast Region and Beyond. NAS Press, 2024.

[10] Dannenberg, Andrew L. et al. Managed retreat as a strategy for climate change adaptation in small communities: public health implications. Springer Nature, 2019.

[11] National Institute of Building Sciences. Natural Hazard Mitigation Saves: An Independent Study to Assess the Future Savings from Mitigation Activities. NIBS, 2005.

[12] International Panel on Climate Change Reports. Sixth Assessment Report. IPCC, 2023.

[13] Snyder, Blake. Save the Cat. McNaughton and Bunn Inc., 2005.

to compete with this research. There is no Troy Beach, Valley Subdivision or Hilltop. Inspiration for this story came from the Coastal Tribes and Nations of Washington State—the Hoh, Quileute Tribe, Shoalwater Bay, and Makah tribes, and the Quinault Indian Nation, all of which are in the process of moving to high ground in response to newly understood tsunami risks and climate threats.

Resilient stories can:

- Resolve internal conflicts, and
- Be shared among friends and colleagues, building trust, friendship, and identifying common ground.

Resilient Shores can be read as a success story of how a community moves to safe ground. It is also an instructional guide for inspired readers to help them create or recreate aspirational stories highlighting more realistic futures. Finally, it's a story of how a public process can be fashioned to build trust and raise issues supporting their quest for resilience.

As a story of Troy Beach, you will enjoy a mystery wherein Watson and his friend Sherl describe how the city came to address climate threats and how the neighboring Hilltop community benefited from their retreat.

As a guide, we welcome you to create your own story for a subject of your concern. Your story may be a home, store, workplace, or city. Your story can progress chapter by chapter, or you may wish to read all chapters and then go back to begin your story. Regardless of whether you create a "Who did it?" or, more aptly, a "In the future, It won't be done again?" story, enjoy the process.

Read as a mystery, *Resilient Shores* is not WHO-dunit, but HOW was it done? How success can occur when so many don't even see the need? Don't recognize the threat.

Our story of Saving Troy Beach's reveal is "What can be done so that as IT happens, the most severe adverse impacts can made tolerable."

To make our story of *Resilient Shores* interesting and provide guidance should you wish to create your own story, it more or less tracks the Freytag Pyramid approach to storytelling and encourages you, the reader, to create your own stories following a similar approach. A who, what, where, when, why, and how method is offered.

Reflecting the Freytag pyramid,[14] Chapters 1 through 2 provide the exposition, with Chap. 1 presenting the disaster incident and Chap. 2 the realized wish, the WHAT and WHERE that emerged. The following 2 Chapters raise the action, our plot, and present how, and why this realization was necessary. Our characters, the WHOs, introduce their perceived risks and aspirations as they tell their stories.

Chapters 5 and 6 provide the climax and resolution. Chap. 5 expands on realized Goals and Chap. 6 presents the HOW and WHY.

[14] Writers.com. Freytag Pyramid.

The action falls and our story could end there, but it is important to view the resolutions in context. Chap. 7 provides additional procedural detail and Chap. 8, context to the resolution, completing the story's HOW, and WHY.

To support your desire to create achievable stories that will better implement your wishes, *Resilient Shores* tracks an approach offered by Gabriele Oettingen.[15] She is a behavioral psychologist whose research supports exploiting the advantages of positive thinking and has proven effective in achieving desired results. Dr. Oettingen suggests that quests begin with wishing and envisioning desired futures, only the seeker cannot end here. She cautions, however, that dreams may be emotionally fulfilling, but they can also function as ends in and of themselves. These musings can reduce blood pressure and in doing so, lower energy levels. However, here lies a weakness. Visions can function as ends in and of themselves. She recommends that dreamers confront outcomes to their wishes, and face obstacles—face the real world and formulate achievable plans. Her research centers on individuals, but we feel the process offers similar opportunities for communities of individuals seeking common ground. Collective action.

Accordingly, Chap. 2 introduces the realized Wish, while Chap. 1 provides context. Chapters 3 and 4 introduce Mental Contrasting. Chap. 5 reintroduces the achieved goal and more. Chap. 6 reveals HOW the wish was accomplished, and Chap. 7 expands this explanation. Chap. 8 critiques *Resilient Shores as an* optimistic, realistic, and achievable story. The book concludes with a "What Next" discussion. Here, Cassandra offers suggestions for initiating a storytelling planning process within your communities.

As was mentioned earlier, *Resilient Shores* can be read in three ways: as an interested observer, as a guide for personal action, or as a manual for inspiring collective change. It offers an opportunity to confront conflicts. To clarify outcomes and the obstacles to your wishes. To create a story —complete with rising and falling action. To devise compelling characters, resolve the challenges you present. To build a story. A story that not only guides you toward a new path but also one you can share with friends. Stories that uncover common ground and spark communal action.

Whether you read *Resilient Shores* as a mystery to fill an otherwise uninteresting interlude, or to help write your own "What Can Be Done in My Community" story, we hope the process will help you to confront this challenging subject—as a class, study group, book club, community group, family, or as a questioning individual.

Whether you or your community faces rising seas, stronger and more frequent storms, or unbearable heat, tackling climate challenges by creating, sharing, and reshaping your stories can result in real change. Creating and telling stories can allow you to assess threats to what you value and suggest a host of risk-reducing approaches. Even the most frightening conditions, when confronted as stories, can address financial, emotional, social, and spiritual conflicts, safely. They are just stories after all, and resolving conflicts through storytelling is something we've been doing naturally our whole lives. Let's put it to good use.

---

[15] Oettingen, Gabriele. Positive Thinking Inside the New Science of Motivation. Penguin Random House, 2025.

If you decide to create our story while reading *Resilient Shores*, don't get overly macabre; the exercise won't be fun, and no one will listen to you for very long. The characters you will be introduced to while reading were not necessarily elated with having to address a changing climate, but they confronted the need without unfettered duress.

Think of our changing climate as a problem we all have to address. There is no benefit in ignoring it. Like crossing a street, we just have to get to the other side without getting hurt. Creative problem-solving can be fun, and like storytelling, as we have said many times, we have been doing it all of our lives. It is a skill we teach our children and grandchildren.

While reading, while creating your own climate resilient story individually or with friends and families, add a bit of whimsy, and have fun with the exercise.

Your stories could be personal, aspirational musings, or directed to larger audiences. To friends, neighbors, and colleagues. You could tell your stories as grandparents, advancing the wishes of your grandchildren. These younger audiences, our future generations, should not be sidelined. Tales of stranded dogs and rescued puppies may capture their interest. Other children may be drawn in with accounts rising waves slamming against unprotected "butts." Having "farts" propel them to safe ground, may capture the dear younger soul's attention. At least for a few seconds. And maybe for a lifetime.

Rising seas, hotter weather, and more severe storms will change all of our lives.

Enjoy every moment of your reading, musing, creating, wishing, planning, and achieving success through storytelling. Your storytelling musings, your journey, will occur in a safe space. Don't hold back. Be honest with yourself, and let your imagination run free to see where it leads for you and your community.

Now on to Resilient Shores, the story.

Contents

Part I
Exposition

Chapter 1
Hurricane Apollo

Values and assets *introduced*. The challenge—defining the conflict and how to resolve it?

Abstract Once upon a time…. There was a loud crack. Lightning severed the old oak tree. Blown across Front Street, it rammed through Paige's café window. Storm-driven seawater followed, and the café's interior became a scramble of chairs, tables, and shelving. *Resilient Shores* begins with the arrival of Hurricane Apollo. Characters are introduced along with their fears and aspirations. You, the reader, are invited to create your change agent, the threats, risks driving your story, along with those impacted.

Keywords Exposition · Hurricane · Storm surge · Freytag's pyramid · 5 Ws + H · Risk · Phases in emergency management

Troy Beach Setting—Today

Artist: Ron Kasprisin

Energy transformed, not lost

Once upon a time….

There was a loud crack. Lightning severed the old oak tree. Blown across Front Street, it rammed through Paige's café window. Storm-driven seawater followed, and the café's interior became a scramble of chairs, tables, and shelving.

The residents of Troy Beach knew Hurricane Apollo was coming. Many followed its path when it began in the mid-eastern Atlantic Ocean.

B. Freitag, *Resilient Shores: Adapting to Climate Change Through Storytelling*,
https://doi.org/10.1007/978-3-032-09047-8_1

Cassandra, the City Planning Director, had posted its track on the city's website. Still, many were not aware of the threat. After all, it was only the beginning of hurricane season.

Hurricanes had brought a familiar sequence. More beach sand would wash away, the coastline would retreat farther shoreward, and all impacted by the storm would begin an all-familiar path to recovery.

Before Apollo hit the shores of Troy Beach, those few watched every slow-moving mile it took, undoubtedly wondering if it would make a direct hit on their small city.

It did.

When it made landfall, Category 4 storm-forced winds reached 160 miles an hour, and waves, over ten feet high, washed onto the city's commercial area.

The animals across the region didn't have to be told. Birds flew inland, snakes slithered under cover, rabbits hopped into protective brambles, and any living creature that could move tried to avoid the storm's path, even before the howling winds came. This was hurricane country, and through thousands of years of coexisting with the climate, their DNA seemed to carry the message of "flee" when a steep drop in barometric pressure combined with the increase in wind. They didn't know why but knew it was time to move out. Darwin's survival of the fittest had worked over millions of generations to produce animals who had this particular instinct fitted to this particular area.

The newer zoological arrivals—horses, cows, family pets—fussed with anxiety, feeling the strange weather. Besides, their people were acting in strange ways—some packing to leave, some boarding up windows. The luckier animals left with their families before the hurricane landed, going far enough away to miss the brunt of the storm.

Hurricane Apollo made landfall, and the first lashes of wind and driving surf hit the shore well ahead of the eyewall that packed the heaviest wallop. Rains preceding the violent winds. Tornados spun off from the main storm. The roar gradually became overwhelming. The seas crashed onto the shore and through homes. The sky raged, and trees whipped around, stripping away their foliage. At first, it smelled of salt from the surf. But as buildings tore, trees shattered, and pipes broke, the smell darkened into a stench, reflecting a failing human settlement that ocean water could not scrub clean.

Winds stripped the dunes of sand, cutting away at the seagrasses and creating corridors through the protective citadels. Wind-driven sands etched their signature on walls and tree trunks. Storm surges rushed through these openings. The dunes eventually disappeared, only to re-form inland, behind beachfront development.

Thunder and lightning competed with the frightened screams of people who stayed behind and didn't realize, or underestimated, what awaited them. Not a carnival ride but a day and night that would change their lives.

The Admiral, as the Harbor Master was affectionately called, wanted to be home with his family but knew his responsibility. While the storm was approaching, he went to the port to save the recreational and commercial fishing vessels. Many of the fishermen came with him, trying to save their boats. The wind, rain, and ocean beat them relentlessly for hours. The fishermen realized there was nothing else to be done at some point, and they obeyed the city's warning to leave the premises.

The port had been enhanced with a seawall, and it held back the surge—for a while. When it was overtopped, just one vessel breaking loose started a chain reaction, washing the recreational boats into the fishing fleet moorage. The heroic efforts of the Harbor Master and the fishermen were washed away as Apollo destroyed boat after boat after boat that day. The punishing beating of the sand and debris against the pavement and structures also severely eroded the boat ramp.

Along the shore were multistory condos. Sandwiched between and behind these monoliths were single-family beach homes. Hurricane Apollo didn't let a few buildings contain its strength. Some of the best-built structures—the newest ones mostly, built to strong building codes—survived intact. Just a few.

Waves and storm surge undercut the supporting foundations of all the beachfront structures. The newer homes did better, but all were weakened, whether elevated atop high piers or protected by intervening structures. Some toppled under the relentless surge and wave energy. Others blew apart as the wind's force on structural elements crossed safety thresholds. A few endured as a compromise. Their foundations stood like giant tables, with their living areas gone—distributed as debris throughout the neighborhood. The airborne debris became missiles during the storm, plummeting and destroying structures that could otherwise have survived. Streets and utility systems were not spared. Electricity, water, and sewer services stopped for days, weeks, and months after Apollo came to call, leaving the people of Troy to fend for themselves, reminiscent of their forefathers.

All people should have left the shore. Those who could go inland and stay with relatives or take a chance to find a place to stay for a few days were gone. But there were newcomers, tourists who had no idea what this hurricane had in store for them, and a few recalcitrant homeowners.

Gadsden, one of those homeowners, and his friends held a hurricane party on the balcony facing the beach. They'd been warned of the danger several times by police and firefighters. Finally, they were told to write their names and contact info on their arms so they could be identified by their next of kin. The first responders warned no rescue efforts could be made during the brunt of the wind and flooding. If—no, when—the partygoers met with trouble, they were on their own until it was safe for rescuers to look for them.

By the time Apollo made landfall, Gadsden and his partygoers had retreated to the strongest room in the house, the bathroom. They had left the Great Room, the centerpiece of which was a large picture window facing the water. The room extended to the shoreside of the house where he had installed a second, albeit smaller, picture window. During the height of the storm, debris from a neighbor's home blew through the large window, crossing through the living area and out the back of the room through the smaller window. Wind and wind-driven surges exceeded the living area floor height, but the forces did not destroy the structure because the determined floodwater had an open pathway through the building. The bathroom, along with its occupants, survived.

Others were not so fortunate. A family who was excited to vacation along the shore rented an adjacent condominium. They heard the warning but didn't comprehend the danger. Thatcher, the developer, had her contractor Atlas install a warning system in each apartment. The device worked, and a siren screeched out its warning. The vacationers heard the signal but couldn't interpret the warning. They didn't know how dangerous a hurricane could be. Their building looked strong, so they found a way to turn off the siren. Being on a higher floor, the winds were more substantial. When the balcony furniture broke loose and slammed through the sliding door, these guests had little time to find cover. The parents were frantically searching for their children and happened to be in an interior hallway, and they lived. The children were not so lucky and died after being struck by airborne debris. Their tragedy continued after the storm when the parents were unable to forgive themselves and could not maintain their marriage.

Farther up the shore, the wide sandy beach narrowed, running up against a small rise of about six feet. It was appropriately called "The Unhappy Bluff" by everyone in town. At one time, a large Loblolly pine grove offered protection to the hillside against wind and surf. The grove had been reduced along with the size of the sandy area. The remaining sliver of a beach was once broader, but engineering attempts at armoring the embankment redirected the wave energy, eroding The Unhappy Bluff and increasing the inland shore migration.

Hurricane Apollo took another hundred feet of shoreline and further reduced the size of the protective forest.

This area had once been a plantation and was still owned by that family. The current occupant was Isabelle, or "Belle" as she preferred, Troy Beach's oldest permanent resident. Belle had fallen on hard times and had subdivided much of her ancestral land. The new subdivision owners were wealthy out-of-towners who each attempted to recreate similar plantation-style homes. These new homes were built closer to the beach. All of these new transplants had reduced the size of their coastal protective forest. Water rushed around these newer homes, scouring foundations and further eroding the protective rise.

Front Street was behind the beach and shorefront housing, which ran parallel to the shore through most of the Historic Downtown District. The wind continued its direct and constant campaign here, and accompanying surges continued their destructive campaign. The storm surge washed over most of Front Street but waned considerably as it lost energy when it reached the slightly higher elevation in the back of the district.

Prominently sited along Front Street was the Troy Beach Café. The owner, Paige, had evacuated when warned. She also ran a food truck that she drove to a safe location. She was fine. The Café was not.

Storm surge broke through a doorway; saltwater washed sand and debris into the café. Water dismantled the appliances, and the furniture became battering rams, further destroying the café interior. Gas pipes ruptured, but the wind and walls torn open prevented gas from pooling, so there was no explosion or fire.

Down the street from Paige's café was the "Bargain Beach Wear" store. Cap, the store owner, had evacuated. Windows and walls broke from hurricane gales, though seawater didn't enter the building. Much of the store's merchandise blew onto the street. The store's name was evident from street level, where the store's branded bucket hats, covered in mud and leaves, could be seen piled against the building. Hard rains accompanied the storm and horizontal torrents of water-soaked display items inside the store.

The Front Street concrete corridor had buckled in places. The storm surge had washed away the pavement. This, along with debris deposited haphazardly on its surface, made driving on the street impossible.

Behind Front Street and away from the beach were the city's more affordable, moderately sized homes. The storm surge did not reach these neighborhoods, but winds and flying debris caused extensive damage. Like the rest of the city, these residents lacked city power, water, and sewage service after the storm. Deficient of resources before the storm, rebuilding was harder for the people in this neighborhood. Some of them just moved away, unable to either afford recovery or face the thought of another experience like Apollo.

During the storm, the lack of a speedy medical response contributed to the hardship.

A recent resident was severely injured when his wife asked him to go outside and retrieve their prized dog, prized more for being cuddly and submissive than for his self-reliance. The male Miniature Dalmatian bolted when a neighbor's picnic table was blown airborne and slammed through their front door, and scampered away. The family lived back from the beach, shoreward of the storm surge, but not protected from hurricane-force winds. The obedient husband's trauma was not the direct result of his initial wound. His initial superficial injury worsened with time, as rescue workers could not reach him until the storm passed.

He was lucky and survived. Another neighbor ran outside during the height of the storm, looking for her pedigree cocker spaniel, when a piece of debris hit and killed her.

All in all, four people died in Troy Beach that day. Many more were injured, both physically and emotionally. Apollo left a scar on everyone in some way.

The community of Hilltop overlooked Troy Beach. In Hilltop, residents weren't worried about surges, rising seas, or waves, but Apollo still brought respect. Winds and rain blew roofs apart and soaked living areas. Repairs would be expensive, and many people couldn't afford insurance. At the suggestion of Angela, the young Music Director, all were invited to gather in the Hilltop National Baptist Church, the safest structure in the neighborhood. Ozell, a neighborhood handyman, had helped Angela distribute preparedness information before the storm and now helped make people comfortable when the Eye passed.

When Hurricane Apollo roared through, it was a sound they'd heard before. Hurricanes were becoming more frequent. The older resident families had lived in the area for generations and were familiar with the routine—board up, move fragile items to the bathtub, stay sheltered inside.

After relocating the fishing fleet, reorganizing the beach, tearing through the shorefront homes, stopping commerce along Front Street, and soaking Hilltop, the Valley Subdivision was Apollo's next victim. It was not on the beach, so it missed the direct force of the waves, but the water still did tremendous damage. Storm surges were blown up the Valley River, producing a wall of water and preventing river flows from draining the punishing amounts of rain. This blockage caused floodwaters to flow onto the Valley Subdivision, or Valley as the community was called.

Floods came in several pulses. Eventually, the storm surges lessened and receded. As these first flows began to recede, rain pounded the upper Valley River watershed and drained onto the subdivision. Apollo seemed to rest over the watershed, and rains continued for several hours, contributing to the area's flooding. High-velocity flows were confined to the channel, but the excess water flowed onto the natural floodplain on which the subdivision was built.

The buildings did not escape wind damage either. The valley topography directed gusts through the subdivision, causing further destruction.

The community was fortunate to have a resident dedicated to storm preparedness. Ezra was born and raised in the area. He led the local response organization, the Volunteers Active in Disasters, or VOAD, and held preparedness exercises. Being active in the Valley Southern Baptist Church, he had an extensive network of local volunteers to call on.

Much of the community lay within the FEMA floodplain and often flooded. However, the flooding over the flat river terrace was usually only a few feet deep. Ezra and his volunteers had distributed sandbags and preparedness kits when a hurricane watch was announced, and they were prepared to offer assistance as soon as the floodwaters receded. Mold would be a special threat then. Following the last storm a child died from mold complications, and Ezra had been working on a quick response team specifically trained to address mold.

Lastly, there was a small community asset located near Hilltop. Nestled in a forested depression, high above the shore, and protected from violent winds was a motel that provided a safe haven for the community. Locals referred to it merely by the owner's name and called it Raj's. Raj's real name was Rajanikant. I believe Kumar was his surname. Rumor had it that he came to America as a servant, although he was very protective of his past. During the storm, it sheltered those who could not reach or didn't feel comfortable evacuating to the National Baptist Church. After the storm, it offered a place for those without power to cook, take showers, or merely gain emotional strength. All three communities had come to rely on Raj's for support.

Apollo made landfall and wrought destruction. This time was different. It also brought an opportunity.

Cassandra had envisioned a new, safer future for her city. She knew changing direction would not be easy, but she was determined to begin the process. Her most vocal and outwardly confirmed ally seemed to be Allison, a resolute high school student.

As for the runaway dogs, after a few weeks, a more devious and self-reliant dalmatian returned home, accompanied by a cocker spaniel. Two months later, Troy Beach was blessed with a litter of exceptionally cute, fluffy-eared spotted puppies. Troy Beach is bouncing back.

Your Own Story: Cassandra's Suggestions

Problem/Exposition: Values and assets, *introduced.* The challenge—Defining the conflict and HOW to resolve it?

The story plot problem, our crisis, is introduced, but HOW will it be resolved?

By way of an introduction, Watson has asked me to help you create your own resilience story. My name is Cassandra, and in case you didn't pick this up while reading Chapter 1, I was the Planning Director for Troy Beach when Hurricane Apollo made landfall. Watson and I know that many people feel they live in places subject to increasing climate change-related threats and may benefit by thinking of disasters and disaster recovery differently. We feel that in creating resilience stories, you may be able to put a face to your problems and come up with helpful resolutions.

Developing your own story may help you discuss issues that are too difficult to address otherwise. Your resiliency stories may function as action plans. The objective here, is to provide a way, through storytelling, where you can confront the seemingly impossible, safely.

At the end of these chapter exercises, you will have the opportunity to create a revised vision, a story presenting a new future for you or your community, and in the process, a realistic implementation plan.

Exposition/Conflict

Are you starting story with an event, after or before?

Have you introduce the Whos, Whats, Wheres, and Whens needed for your reader/ listeners to understand the context.

Your exposition will form the foundation from which you will propose probable resilient futures in Chap. 2. The conflict may involve HOW accomplishments occurred, not so much WHAT was completed.

If you are interested in creating your personal resilience story, I welcome you to follow along with the suggestions I have included at the conclusion of each chapter. I have also included some supporting reading. If undertaking this exercise does not excite you, feel free to go straight on to the next chapter.

Before you dive in, you might want to check out the workshops we ran after the storm, written up as a screenplay in Appendix A. Here's a quick recap:

> After Hurricane Apollo, I was tired of the same old cycle—build and rebuild after every storm. Most of Troy Beach wanted to return to life as normal, but I wanted to explore new ways to bounce back. So, I organized three workshops to help the community take charge.
>
> The first workshop focused on identifying what we valued in Troy Beach and the assets supporting those values. The second was about understanding how and why these values were at risk. The final workshop encouraged brainstorming strategies to protect those values. These workshops were fun, self-directed, and built trust within the community. We also aimed to develop clear ideas that planners could turn into long-term strategies.
>
> If you're still interested in creating your own resilience story, read on.

5 Ws + H

"I KEEP six honest serving-men (They taught me all I knew); Their names are **What** and **Why** and **When** And **How** and **Where** and **Who**. I send them over land and sea, I send them east and west; But after they have worked for me, I give them all a rest."

The Elephant's Child, Just *So SStories*, udyard Kipling, 1902.

Fig. 1.1 Six fundamental questions

I will be offering detailed guidance at the end of each chapter. Some will be questions to ask yourself, but more complete suggestions will be on:

- **Plot Development**. Stories have beginnings, middles, and ends. Real life, not so much. Think of any real event, and you could probably come up with dozens of ways to tell what happened.
- **Resilience Concepts**. *Resilience Shores* is a story of combating climate change. Watson and I do not provide much information on climate science; that will be up to you, but we do offer guidance on resilience thinking.
- **Conflict Resolution:** Here, I will discuss how our characters, what might be thought of as disaster archetypes, define, assess, and resolve conflict. Become resilient.
- **Paige's Story**. Paige's evolving Resilience Story Board will be offered as an example of how you can take advantage of this tool in creating your story(ies).

Plot Development

Most of you have heard of the 5 W + H storytelling model—Who, What, When, Where, Why, and How. Rudyard Kipling popularized this timeless approach, and it is as good as any, so we'll use it here. (Fig. 1.1).

As for your plot, consider shaping these 5 elements within the context of the Freytag story pyramid, a classic guide since 1863. It ensures your story has a beginning, middle, and end, building suspense and intrigue along the way. For Troy Beach, this means presenting the Who, What, Where, When, and How while transitioning from exposition to an actionable climax, and finally, resolution. Chapter 1 sets the stage by introducing the values and assets under threat. Future chapters, shown in softer hues in Fig. 1.2, help outline the full story, which can evolve as you go.

Let's begin the exposition by addressing the Who, What, When, Where, Why, and How.

Resilience Concepts/Conflict Resolution

Who is involved? There are several WHO's you'll need to determine. WHO will tell the story, WHO is the story about, and WHO is driving the plot? You, of course, will be one of your WHOs.

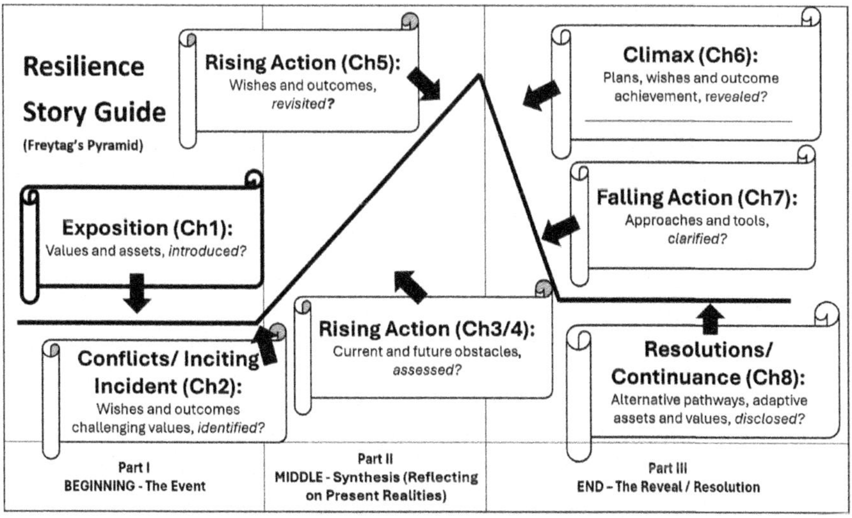

Fig. 1.2 Freytag's pyramid—exposition defined. Prepared by Bob Freitag

In our story, Watson is, for the most part, the storyteller. He tells the story in the first person. You, too, could create a narrator to tell your story. You could also tell your story in an anonymous third person.

Or, for your story, you could ask a lot of questions and have the listener tell the story as a Questions and Answers, but this may be difficult.

As for WHO the story is about, Watson has our typical community resident characters tell their personal stories by describing their values and fears. These individual stories drive the plot and culminate in a vision for a resilient future that supports commonly held values. Watson wants you to understand that corrective actions and hazard mitigation measures are personal. Enjoy creating your characters. I am sure some will be based on real friends, and some you will make up to reinforce some themes or lighten the mood by telling a joke.

You will discover as you read that not all are happy with the outcome. A resolution that may be resilient for one of your characters may cause distress for others. Watson confronts this dilemma in Chap. 8.

Our Troy Beach characters were created to introduce you to a broad range of climate-related concerns. The characters represent disaster archetypes, and you will probably recognize them as disaster archetypes. You will meet homeowners who feel they have lost everything, the recovering owners of damaged businesses, conflicted churchgoers confronting irreconcilable change, and the guardians of threatened traditional values. We hope their stories can help you discuss and build your stories of resilient futures.

Feel free to experiment with different storytellers. Devise different points of view. Some characters will make their appearance only after you are well into your tale, to clarify a detail in your plot. Or add some humor. Others will be introduced at the beginning driving your story arc or tying up loose ends. I will discuss the importance of a story arc shortly.

WHAT will your story be about? For *Resilient Shores* at one level, it is about a community trying to adapt to change. It is also about individuals such as Paige who is trying to maintain her lifestyle, the Admiral managing the Port, and Gadsden being allowed his independence. Herein lies conflict. Can they all get what they want? Resilience is personal.

Stories are all about resolving conflict, and what has more opportunity for resolving conflict than individuals comfortable with their existing life, being forced to change?

WHAT challenges or problems can your characters face and resolve? It could be anything from unraveling why shores are receding to how to adapt to the change, to overcoming personal obstacles. The WHAT drives the plot.

You also need to think of the story arc at this point—the journey your plot takes from beginning to end. The WHAT you introduce at the start of your story, must not be forgotten. Loose ends must be tied up as you conclude your conclusion. The WHAT you introduce here, threats to your character's values, for instance, must be resolved at the end.

An outline may help.

This chapter introduces you to Paige and Cap. Their businesses have been destroyed. Watson must describe how the threats to their values and supporting assets can be resolved. Will they repair their business, leave town, change direction…?

Do not worry if you do not know the exact ending at this point in your story. Often, authors resolve conflicts only as they develop their stories. The ending can be as much of a surprise to an author while writing as it will be to their readers. That's the fun part of story writing—alone or in groups. The storyteller, as she develops her story, often not unlike the reader, discovers interesting characters and conflicts as the story evolves. As it is retold.

The WHAT also refers to threats. What is causing the "disturbance in the Force?" For us, it relates to hazards impacted by our changing climate.

I find it best to think of hazards as having four hazard-profiling characteristics—Intensity, Location, Frequency, and Timing.

- **Strength/Intensity.** The hurricane Apollo brought winds of over 100 miles an hour.
- **Location.** It hit Troy Beach differently—seawater swept through downtown, freshwater flooded the Valley, and winds wreaked havoc in Hilltop.
- **Frequency.** This wasn't seen as a one-off; people recognized the growing frequency of severe storms.
- **Timing.** The storm wasn't sudden—there was time for preventive measures and evacuations.

As for impacts, I group them into three asset types: Built, Natural, and Social.

- **Built Assets**. The built marina was severely damaged, as were commercial establishments along Front Street.

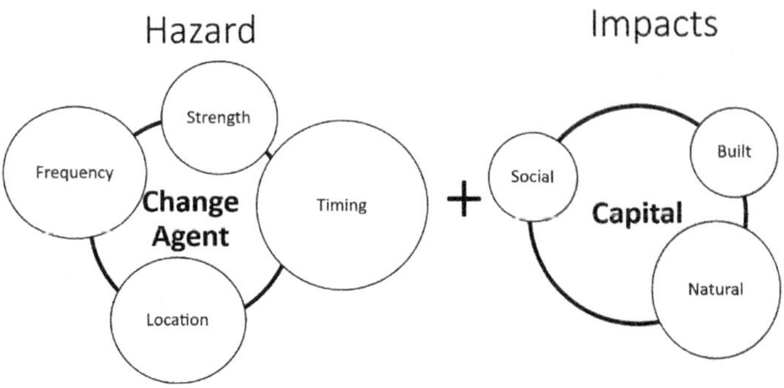

Fig. 1.3 Risk defined. Prepared by Bob Freitag

- **Natural Assets**. Natural dunes and bluffs were eroded.
- **Social Assets**. For social network asset impacts, residents had difficulty cooking food, but not all was bad. Friendships were strengthened, as were a host of mutual aid networks. Trust was increased. Eventually.

Risk is usually thought of as the relationship between hazards and impacts[1] (Fig. 1.3).

WHEN will your story take place? Our story of Troy Beach is presented in two time periods. The first setting addresses the present. This is what you just read. The hurricane event has just occurred, and our characters are responding to Hurricane Apollo. They are reflecting on their recovery. This will be discussed in the next chapter.

You will need to determine when your story begins. Thinking of Emergency Management actions as having three phases[2] can help—Measures that reduce risks before an event (Preparedness), during an event (Response), and following an event. Our Troy Beach story begins with the event or the Response phase.[3] Most types of disasters re-occur. They are best thought of as part of a cycle (Fig. 1.4).

In other words, you will have to determine WHEN the WHAT you introduced in your story takes place. Does it take place during the event, during the Response phase? Is it about Recovery, following the event, or is it the setting before the next event, Preparedness? Or does it weave in all three?

WHERE does your story take place? For our story, there are three distinct key locations. Troy Beach proper, the Hilltop community, and the Valley Subdivision. Each will allow for a discussion of alternative resilience stories in the pursuit of commonly held values. There are also places of interest within these locations. For instance, the marina is included within the City of Troy Beach; Troy Beach, and Hilltop both have historic districts.

---

[1] Freitag et al. A New Approach for a New Era, Ch 2. Island Press, 2009.

[2] FEMA. National Incident Management System. Third Edition, October, 2017.

[3] FEMA. Mitigation and the Emergency Management Cycle.

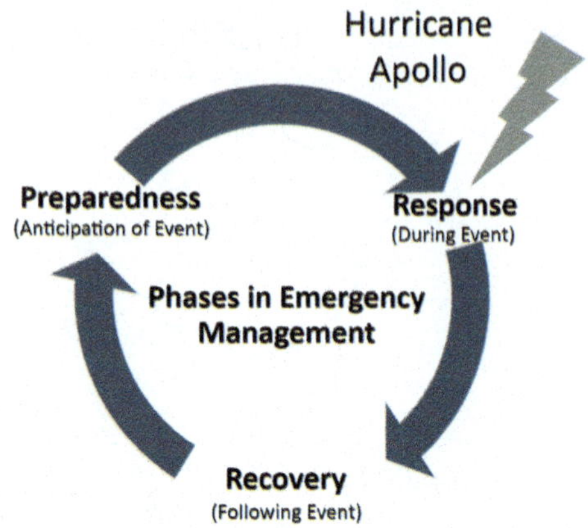

Fig. 1.4 Phases in emergency management prepared by Bob Freitag

As with building your characters and defining the WHAT causing the conflict, you may need to develop your story to discover all the places of influence.

The **WHY,** is the conflict? The story is not just about what Troy Beach, the city, or the communities of Hilltop or Valley became. You will discover what they become in the next chapter. The WHY will be revealed as the story progresses, as will the HOW. The HOW and a little WHY drove most conflict.

Our characters' struggles revolve around how long-term environmental threats to their values and associated assets conflict with WHY these communities cannot bounce back to a normal to which they are familiar.

HOW the threats will be resolved also contributes to conflict. Watson can be thought of as a detective who solves a crime. Climate Change is the criminal, the disaster event, the crime. Troy Beach the crime scene. The HOWs are the resolution and the result of Watson's sleuthing. The WHY is revealed as the story progresses, and there are as many WHYs as there are characters. Remember, resilience is personal.

Are you developing a drama, a love story, or even a comedy?

Paige's Story

All genres are possible. Paige could have been written as a detective and the hurricane the destroyer-murder, or she a love story with her café and the friends this relationship brings. Paige could be in some kind of comical, awkward situation where she continually tries to outdo Cap after each event as they both become resilient.

You can start with one genre and change it later. And enjoy the process.

In subsequent chapters, we will weave in WHY your WHAT happened and incrementally discover HOW resilience was being achieved. The Who, What, Where, and When introduced here is the start of your story. This context will more than likely change as you proceed on your path of discovery. Your story will surely evolve.

We have created a template for you to use, and it is included in Appendix E, Templates—Resilience Story Board. It was designed to help you record your ideas and provide a reference as you tell and retell your story to friends and colleagues.

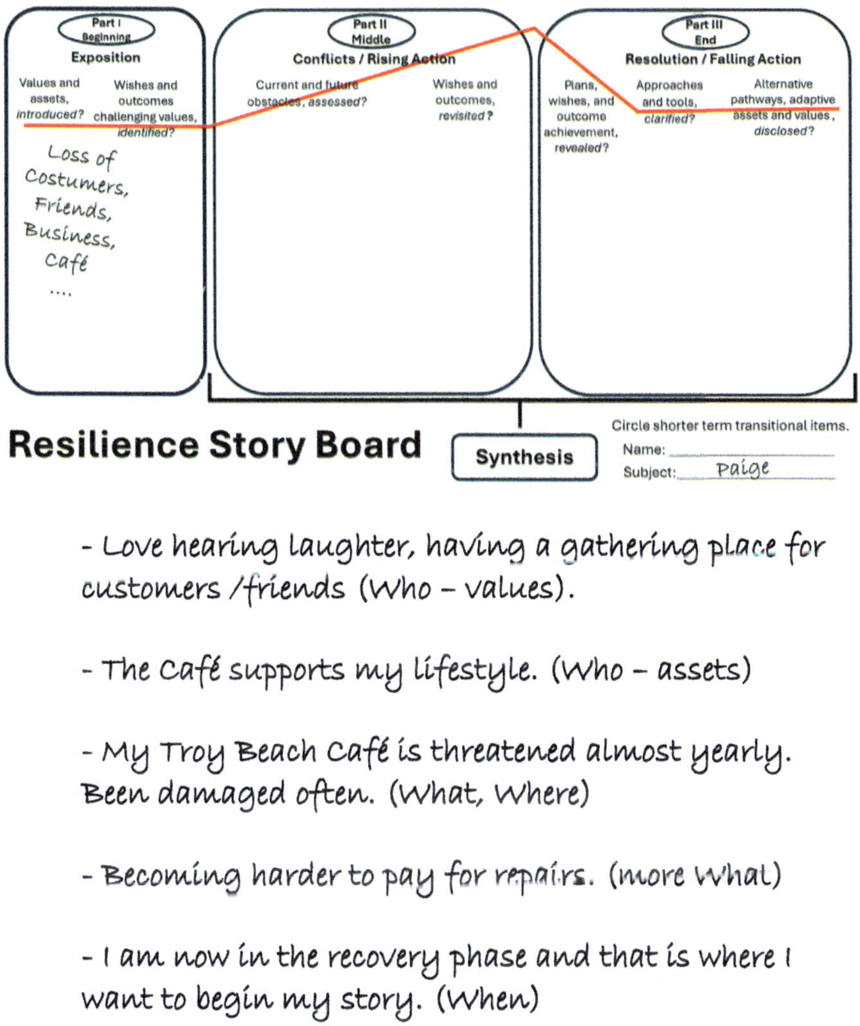

Fig. 1.5 Paige's exposition. Prepared by Bob Freitag

Refer to Fig. 1.5 to get an idea of how Paige may have introduced her story on the Story Board. Her fear is not so much the storm, but not providing a safe place for her friends.

The template has space for you to build your story. It also includes an overlay of Freytag's pyramid to ensure you are reminded to continually build action, and intrigue, maintaining a focus on your resolution. In the "Resilience Story Guide," I have included text reminding you of how we developed our Troy Beach story, but it is a lighter-colored font to help you keep track of your story but reduce its direct influence on yours. The need for synthesis in Parts II and III is highlighted.

Lastly, now that you have been introduced to the cause of our concern, I invite you to learn what transpired shortly after the hurricane and Troy Beach's re-imagining of its future. I held a series of workshops after Hurricane Apollo. Watson will refer to these in his account. These workshops were written as a screenplay to enable you to have more of an active role in the Troy Beach story. You may enjoy reading the screenplay with friends and colleagues or as a play. The screenplay is included in Appendix A.

Chapter 2
Watson Returns

Wishes and outcomes challenging values, *identified*. The challenge—determining aspirations, an end game(s)?

Abstract Twenty-five years after Hurricane Apollo, Watson returns to Troy Beach to chronicle their success in adapting to the changing climate. It had been over 20 years. He marvels at all accomplished and is convinced the tale of Beach's experience could help others. He calls on his old friend Sherl to help. Here, Troy Beach, the realized future vision, is introduced. The WHAT and WHERE for our story emerge. How this New Troy Beach came to be is the question. If you are creating your own story, here is where you may devise a new vision for your community, aspirations, and wishes.

Keywords Conflicts · Recovery · New troy beach · Wishes · Outcomes · Bouncing back · Hazards mitigation · Move

New Troy Beach Setting, Train Station—25 Years in Future

Artist: Ron Kasprisin

What is your/our wish? Our goal?

Twenty-five years after Hurricane Apollo, I came back to Troy Beach to chronicle their strategy for adapting to the changing climate. I'd been there at Ground Zero all those years ago and helped create the contingency plan the city eventually implemented. But it had been over 20 years since I'd left. Now it was clear that other beach cities and towns were having trouble adapting to the new world global

B. Freitag, *Resilient Shores: Adapting to Climate Change Through Storytelling*,
https://doi.org/10.1007/978-3-032-09047-8_2

warming forced upon us. I was sure Troy Beach's experience could help others. I called my old friend Sherl, and we agreed to meet at the Hilltop train station.

Ironically, May 1, the beginning of the revised hurricane season, was a festival celebrating the city's revival. It gave me a perfect opportunity to connect with old friends, witness how the community has changed, and discover our successes—and maybe some failures.

Troy Beach had benefited from Hurricane Apollo by recovering differently. As years passed, severe storms had indeed become more frequent. Higher seas and storm surges forced the shore to move inland. Safety thresholds were crossed, and the city began relocating to higher ground. This elevated land was originally developed and populated largely by the descendants of enslaved people. Most called it Hilltop. It had deep roots dating back to the Civil War and Reconstruction.

I needed to know how the residents of both communities came to accept relocation. What was involved in making this transition?

When I left years ago, many in Troy Beach denied climate change. They preferred to ignore the effects of rising seas, migrating shores, and more frequent severe weather. Hilltop residents were cautious about the need to accept so many newcomers into their already established area.

I started as a student intern with the Troy Beach Planning Department and was assigned to help the then-planning Director, Cassandra, begin her public involvement process. After that, she hired me to stay on as a planner, my first job in the profession.

I was there for the start of the city's reorganization, helped with a series of workshops, and started a new direction for community recovery begun after Hurricane Apollo. I left before most of the implementation started. Today, I was on sabbatical from my university position to write what I hoped would be a book about Troy Beach, Cassandra, and a bit about my own increased awareness over the past decades.

When I began as an intern, I thought the city had little leadership. The Category-four hurricane, Apollo, had smashed through the city a few years before my arrival. The area had stayed in the recovery phase for more than a year—unproductively trying to get back to where they had been. Residents busily gathered up what was left of their lives, trying to rebuild to some familiar normal. I felt their grief, but I knew even back then, that merely using more nails and hurricane ties was not going to help adapt to a changing climate. Some raised their homes a little higher to stay above the next storm surge, but most re-created faithfully what was lost, inadvertently placing them in precisely the same high-risk situation they had recovered from.

I had just returned from being a Peace Corps Volunteer and studied Urban Planning at State. At the time, I weighed in at 135 pounds, and although I was probably as healthy as I would ever be, at 6 feet and gaunt, I could easily go unnoticed. My strength seemed to be, or so I felt back then, listening. I knew little about the planning profession and felt more like an observer. I was to learn that Cassandra was the best I could have observed because she was trying to accomplish something extraordinary.

I hadn't seen Sherl for years and was looking forward to reconnecting with him and exploring together the results of what I had a hand in crafting so many years ago. Sherl had been what I hoped to become. When we met he was older, with over ten years of planning experience when we first met. He wasn't working directly for a public or private planning group but as a community organizer. This increased his credibility with me because my father was also a community advocate with the YMCA, and I valued the role.

Cassandra had invited Sherl to Troy Beach as an observer, an unusual role in a public process back then. None of us realized how revolutionary Cassandra's goals were after Apollo. Sherl was from a river community that was also threatened by a changing climate. Cassandra thought observers from differing backgrounds could bring needed knowledge to our community and provide future allies.

I couldn't see how Sherl could contribute to Troy Beach. Sherl's expertise was rivers, and his hometown hadn't tried any innovative flood mitigation strategies, as far as I could see. (To be fair, they had aggressively used the more proven techniques well, such as levees and low-impact detention ponds.) Sherl often expressed frustration with these recognized approaches he had exhausted back home.

Even so, we became fast friends, and Sherl's evolution with Troy Beach continued well beyond attending the workshops Cassandra brought him to observe. He remained involved as Troy Beach residents' thoughts crystallized into a planning approach accepted by the community and as implementation began.

I took the train from the regional airport, and when I stepped off, I looked around, trying to gain my bearings in a station that didn't exist when I was last in Hilltop. The first thing I noticed was the green. I was overwhelmed by the sheer volume of vegetation. I felt like I was in a very tidy greenhouse, with tiled floors below, benches to sit on, and a coffee shop nestled in one corner. Fruit trees stood along one wall and I identified the scent in the air of oranges mixed with pears. Beneath the canopy of tall trees, shrubs formed another second layer of green. I knew only a few staples–bay laurel, viburnum, and wax myrtle. Vines were hanging from above. Seemingly hundreds of varieties of plants were visible throughout the station.

As I stood, trying to figure out my next move, Sherl appeared.

"Watson, it's great to see you!"

He gave me a double-pump handshake, his left hand closing over our clasped hands. "I was afraid I wouldn't recognize you, but you haven't changed much. You still look a lot like that scrawny kid I remember."

I hardly thought of myself as a "kid," but I guess that's the way it is with people you meet when you're young. Besides, it was Sherl. I didn't really mind.

"As it will always be. I am comfortable being the kid here if, as before, you're comfortable providing your adultish, seasoned, learned perspective," I said, noticing the smile I had almost forgotten.

"How do you like the place?" He gestured around the building.

"It's a bit overwhelming, but everyone else is walking around as if it's nothing," I said.

He laughed. "Welcome to the NEW Troy Beach. This station is as good a symbol as anything else. The beach is still important, but so is being a cutting-edge, energy-efficient community. I won't go into details right now, but the plants help cool the area, provide food, and are partly self-sustaining in terms of water and nutrients. Nothing gets wasted. If it can be recycled, it is."

I must have appeared inattentive while looking around, because I was a little startled when I heard Sherl repeating my name.

"Watson, Watson, What are you thinking?"

"Oh, just wondering what got these communities to attain a singular wish. Reach common ground. Or, if not common ground, safer ground. When I left, each resident seemed to see the problem differently. Their goals, aspirations seemed to cluster, but with little agreement among them. Some wanted to leave, some to develop in a safer place, some want to stay and just gut it out."

"The short answer is probably Cassandra and Raj," Sherl said, "but isn't getting to the HOW, the reason you returned?" The reason you are chronicling our Troy Beach story and writing your book.

"I want to know everything, but first, tell me what you've been doing," I asked. I asked partly because I wanted to know and partly to give myself time to take in everything. A neo-future train station hadn't been on my list of expectations.

We exchanged stories, but I noticed that when I started to walk outside the confines of the stationhouse, Sherl stayed put. I was impatient to see the changes that awaited me.

"Watson, you will see the fruits of your effort today," Sherl told me. "No pun intended." He waved in the general direction of an orange tree. "You will see what Troy Beach is becoming. Later on, we'll meet up with some of the folks you were involved with back in your internship days. Together, we will explore some of the issues that had to be resolved before the city could get to this point."

"Excellent," I said. "Let's go."

"Not so fast, my friend." He put his hand on my arm. "Before you see the new, I want you to remember the old. It will mean more if you remember where we started. Let's get a cup of coffee first."

Building Your Own Story: Cassandra's Suggestions

Conflicts/Inciting Incident: Wishes and outcomes challenging values, *identified.* The challenge—Determining aspirations, an end game(s)?

A resolution is briefly introduced in this chapter. A fulfilled wish. Where the City had bounced to after Apollo, noting conflicts between where Troy Beach was and what it had become. But HOW was an agreement reached?

Here, Watson introduces a realized, more resilient future. Watson has transported you 25 years into the future to a new WHAT. A new normal. In doing so, he has forced you to think of wishes, goals, and trajectories. He has also told you how the story concludes. Values referenced in Chap. 1 are restated here; only many of the assets supporting them have been changed. He demonstrates the conflicts between what was and what the communities had bounced to.

This begins our story and presents the exposition reflected in Freytag's Pyramid (Fig. 2.1).

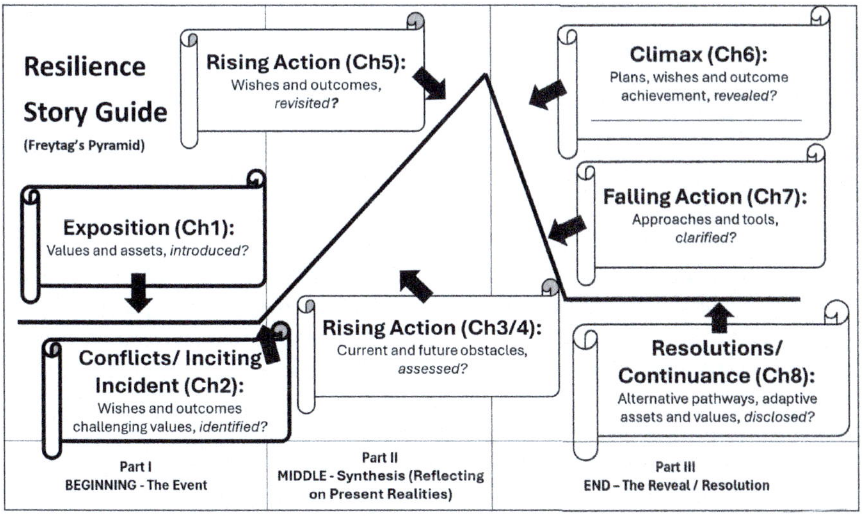

Fig. 2.1 Freytag's pyramid—conflicts introduced. Prepared by Bob Freitag

Exposition/Conflict

Here you will set the stage, introducing the "change" driving your story, those impacted, and the conflicts your story may resolve.

Questions

- Who are the main characters, and what are their values and contributing assets? Their wishes?
- What are the threats and conflicts to these values and aspirations?
- Where and when is the story set? What background information is necessary to understand the story's context?
- Whys and Hows will probably emerge with your story.

At this point, our story introduces threats and offers realized wishes. Conflicts are illustrated. Subsequent chapters will clarify outcomes, confront obstacles, and devise plans, an approach proven successful from the research of Dr. Oettingen.[1] To solve a problem, she supports building on positive thinking practices, where you begin with what you want to achieve, your WISH, then imagine what OUTCOMES might be like when achieving this wish, followed by exploring OBSTACLES to achievement and concluding with creating a conditional PLAN to overcome these identified obstacles. She calls this process WOOP—Wishes, Outcomes, Obstacles,

[1] Oettingen, Gabriele. Positive Thinking Inside the New Science of Motivation. Penguin Random House, 2025.

and Plans. Our *Resilient Shores* story applies the approach to individuals as well as communities of individuals.

Plot Development

For you, following her WOOP process can be the start of your story. Well, probably after you'd introduced some of the antecedent conditions. For Watson, it was Hurricane Apollo.

You can turn the conflicts in your outcomes and the obstacles to your wishes into a narrative. Follow Freytag's pyramid with rising and falling actions. Introduce compelling characters to resolve the challenges you uncover that complete your story arc, and you'll have a compelling story. Stories can be created by a student class, study group, book club, community group, family, or as a concerned individual.

We have provided a template storyboard guide for those interested in writing their own story in Appendix A. This Storyboard Template offers a possible cataloging tool. You may find using a Whiteboard more helpful, along with Sticky notes.

In this chapter, Watson is introduced to the realization of the community's WISH. We will explore outcomes, obstacles, and alternative plans in subsequent chapters. Dr. Oettingen refers to this analysis as "mental contraction," but more about this later.

This chapter completes our "exposition." Key characters have been introduced as has the setting and event that frames the story. The stage is set and the reader, you, now knows that the story is about HOW this reorganized Troy Beach came to be.

In creating your story, you can also suggest obstacles early, or you may let them emerge as the story progresses. There are many ways to tell your story. Remember our story is not driven by WHAT, but by HOW. Just remember The WHAT you introduce at the start of your story must be resolved. You may want to create an outline to make sure your arc is complete.

This chapter presents what Troy Beach has become early so the conclusion is clear and the reader can focus on HOW they got there. It also presents what was imagined, and came to be.

Troy Beach has been made more resilient. Knowing how they became resilient demands an understanding of what resilience means. Resilience thinking can be viewed through different lenses; each may allow you to visualize a unique story arc. A different way you can apply your deductive and synthesis skills. If you do a web search of the word "resilience," you come up with hundreds of definitions. When I did this, I found that these definitions seemed to cluster into three somewhat distinct buckets. Your story arc or plot may apply one or more, with the realization that there is considerable overlap.

Conflict Resolution

Before I introduce these resilience definitional buckets, let's discuss how this chapter developed the five story elements.

Who is driving the story is clarified. Watson is introduced as our narrator, and Sherl is his foil.

What is being seen has substantially changed. Sherl shows Watson what Troy Beach has become. This is the new normal and presents a realized goal.

Where, as a location, remains the same.

When reflects long-term recovery and preparedness. Our Troy Beach communities are in the culminating stages of recovery, and it took 25 years. They are in a better position to prepare for future events.

Why is somewhat clearer, but there are still a lot of questions.

Resilience Concepts

Resilience Lenses. The story being told, and how the new Troy Beach came to be is still only being hinted at. But what is important here, is how the Plot is evolving, and the interweaving of three overlapping resilience approaches—lenses from which to view this story of resilience—Bouncing Back Faster, Bouncing Back Better, and/or Bouncing Back differently. If there is to be recovery, the "bounce" must be to something.

Bouncing back[2]: This definition of resilience has its roots in structural engineering. A structural beam is deemed resilient if it can bounce back to its original condition after being stressed. If the beam is robust, there is little distortion. The speed it can bounce back to its original shape also contributes to its resilience. Used as a metaphor, resilient communities have the ability to bounce back to a pre-event condition. Condition, in this case, refers to the trajectory of "normalcy" of Troy Beach before the hurricane.

In Fig. 2.2, the letter "A" illustrates a community's trajectory. Hurricane Apollo stressed this "normal" condition. If a community creates robust assets or enables its assets to go back online quickly, the community can be said to be resilient. The Troy Beach Comprehensive Plan called for high-rises along the shore. When Hurricane Apollo made landfall, this trajectory was advanced as developers assembled lots and built more beachfront condos. The bouncing back was to a pre-event trajectory.

Think of Gadsden having stored all of the supplies needed to repair his home quickly following a Hurricane.

If the approach fails, however, and communities cannot return to this pre-event trajectory, then the community trajectory may collapse, or at least the values that defined normal may falter.

Bouncing Back Better/Hazard Mitigation[3,4]: This approach is probably the most common. Communities seldom wish to recover to a past vulnerable pre-disaster condition. The disaster exposes vulnerabilities. Communities are also often required by the National Flood Insurance Program and FEMA assistance programs, to incorporate risk reduction measures. It just makes good sense to apply mitigation approaches when rebuilding. Hazard mitigation refers to reducing adverse impacts from a hazard

[2] Environmental Protection Agency (EPA). Regional Resilience Toolkit: 5 Steps to build large scale resilience to National Disasters and Steps to Resilience Video, 2019. https://toolkit.climate.gov/#steps.

[3] FEMA. State Mitigation Planning Key. Topics Bulletin: Risk, Assessment, October 2022.

[4] FEMA. State Mitigation Planning Policy Guide, 2022 (3.2. Hazard Identification and Risk Assessment).

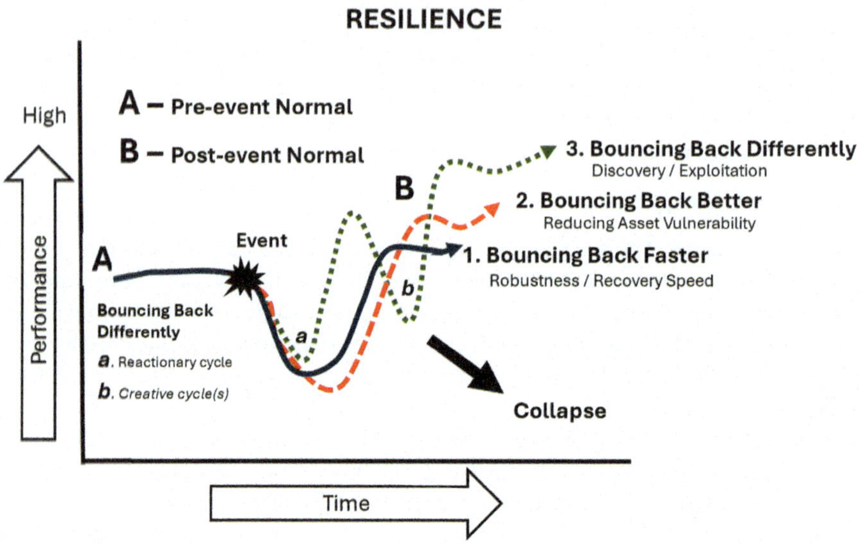

Fig. 2.2 Bouncing back as seen through 3 different lenses. Prepared by Bob Freitag

(flood, wildland fire, earthquake, etc.) and must not be confused with Climate Mitigation, which refers to sequestering greenhouse gases.

Resilience, as viewed through a hazard mitigation lens, focuses on reducing the vulnerability of an asset. In Fig. 2.2, the letter "A" reveals a pre-event "Normal" condition where houses are built at grade. If these structures are vulnerable to flooding because they were built below the flood level, the vulnerability may be able to be corrected by raising the structures above the flood elevation. This Figure shows that when a flooding event crossed some elevation threshold, performance was lost. Houses become flooded and can no longer perform their function. When these houses are elevated, vulnerabilities are mitigated. The letter "B" illustrates in Fig. 2.2, a post-disaster better normal. Individuals or communities of individuals could have also moved off the floodplain or built a protective levee to reduce their vulnerability. Watson refers to these three elements of hazard mitigation as—Move, Accommodation, and Protection, or "MAP" (Fig. 2.3).

Gadsden, the homeowner whom you met in Chap. 1, may be able to reduce his vulnerability through elevation. Paige and Cap need to have customers walk into their businesses. Customers may not be inclined to shop if they have to walk upstairs. Building a protective seawall is presented as a better interim way to reduce vulnerability.

If sea level rise and erosion are also hazards, elevating might not help increase resilience. And, if the elevation process commanded all available capital resources, and there were no reserves for addressing erosion and sea level rise, the community

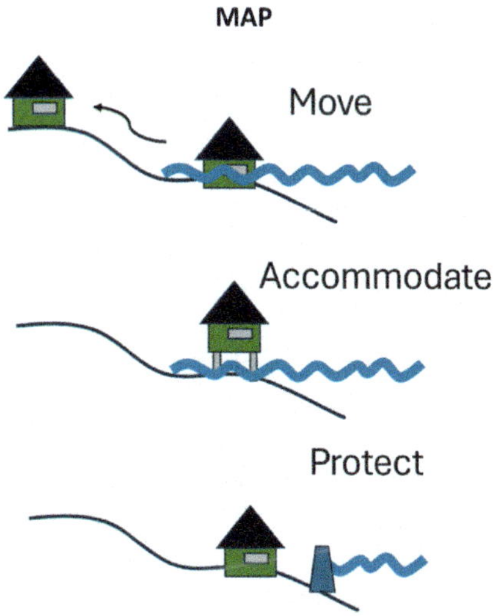

Fig. 2.3 Adaption approaches. Prepared by Bob Freitag

could collapse[5]—the community would no longer be able to support its defining values.

Bouncing Back Differently (Adaptive Cycle)[6]: All organizations go through adaptive cycles as a basic element of their existence. Organizations cycle through periods of growth, conservation, stress, and reorganization. Organizations can also collapse. Figure 2.2, "A" illustrates the pre-disaster trajectory. For instance, Troy Beach was experiencing growth, and its direction seemed stable until Apollo. A changing climate caused events to cross damaging thresholds. Storm frequencies began increasing, and seas began rising. Recovery after storms to the way things were was not occurring.

Bouncing Back Differently can be thought of as having two cycles. The first cycle (a) begins with a destructive event causing uncertainty and weakness, while enabling the exploration of corrective potential and reorganization. The second (b) may enable the creation of a new post-event normal—a new trajectory emerges. Unlike Bouncing Back Faster or Better, this second cycle enables a different pathway to emerge. An organization may loop through this several times before a measure of stability is reached. Or, as with the Bouncing Back Faster or Bouncing Back Better, the system collapses.

Letter "B" illustrates the start of this reorganized alternative pathway, assuring the retention of commonly held values, although many of the supporting assets may be different.

Think of Cap's recovery. Every time his property floods, he repairs it a little better. Accommodating the floodwater with less damage. He may raise the electrical wiring

[5] Diamond, Jared. Collapse: How Societies Choose to Fail or Succeed. Revised Edition Paperback, Penguin Books, 2011.

[6] Gunderson L.H. & Holling C. S. Panarchy: Understanding Transformations in Human and Natural Systems. Island Press, 2002.

while recovering, or store merchandise in containers that can be removed. He may keep all his data backed up off-site.

Paige's Story

The Troy Beach cast of characters views their condition through each of these three lenses. Paige's initial vision is to have protection by constructing a sea wall. This is a bounce back better approach. The application of the lenses you apply will, of course, differ. Your choice will drive your story. You may chase more of a hazard mitigation plot—one where your characters bounce back to a comfortable trajectory. Or your resolution may demand a new path forward where these other approaches are used to create the time needed to reorganize.

In creating your story, you will devise mind experiments, paying with each of these three metaphors or combinations. Your choice will impact your plot/story arc. See Fig. 2.4 for an example of how Paige may document her story using the Story Board template.

If you are using our Resilience Story Board Template, Chap. 1 and this chapter are where you document your problem and possible resolutions.

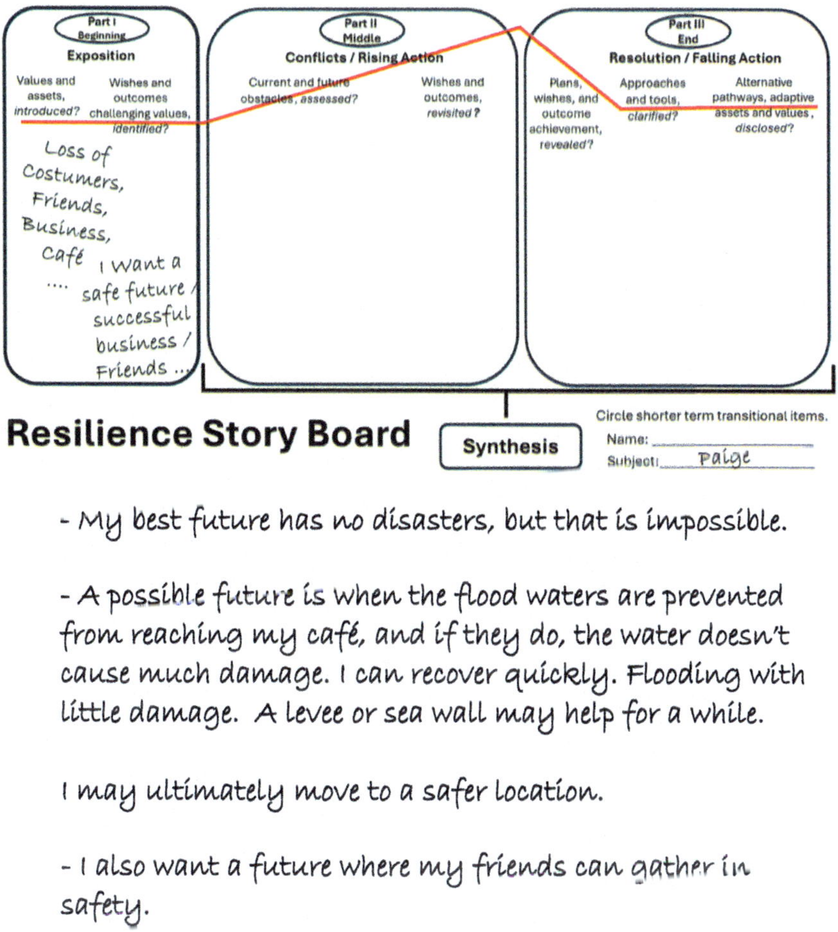

Fig. 2.4 Paige's exposition advances. Prepared by Bob Freitag

Part II
Conflicts/Rising Action

Chapter 3
Sherl's First Visit

Current and future threats, *assessed*. The challenge—confronting present realities

Abstract "Before we explore what Troy Beach is becoming, I'd like to explore what we remember of Troy Beach," Sherl said while turning to Paige's new café. "It will be helpful to recall what the city was like before and shortly after Hurricane Apollo." In this chapter, the action rises, tension builds, and conflicts arise between residents' visions and reality emerge. The WHOs introduce their perceived risks and aspirations as they tell their stories. For those among you who are building their own story, here is where you confront your desired future with some reality. What are the obstacles to your vision?

Keywords Reorganize · Climate change · Housing · Insurance · Permits · Workshops · Obstacles · Agora—Dirty Kitchen · King tides · Strategies

New Troy Beach Setting, Coffee Shop—Remembering Those Recovering from Hurricane Apollo

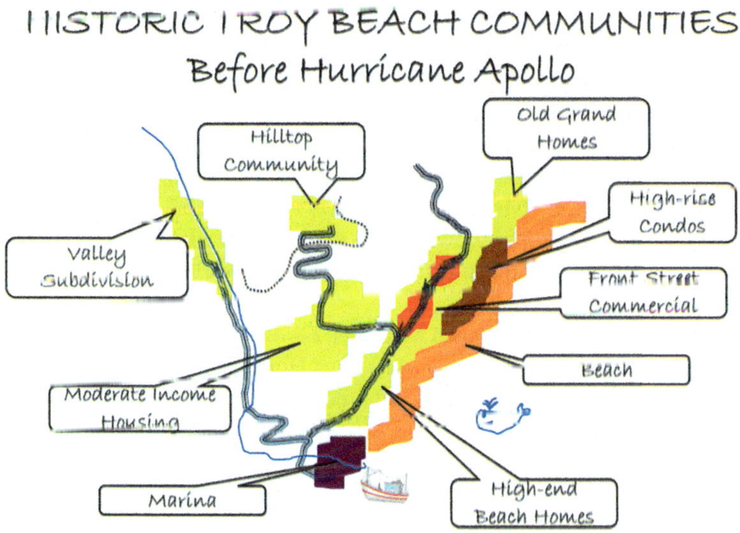

B. Freitag, *Resilient Shores: Adapting to Climate Change Through Storytelling*,
https://doi.org/10.1007/978-3-032-09047-8_3

The location of referenced places. Prepared by Bob Freitag

"Before we explore what Troy Beach is becoming, I'd like to explore what we remember of Troy Beach," Sherl said while turning to a café tucked neatly in a corner of the station. "It will be helpful to recall what the city was like shortly after Hurricane Apollo. Let's have some of Paige's coffee."

He walked to the counter, placed an order, and came back with two cups while I sat at a small table and surveyed the station's use of plants and open spaces. "To your return." He waved his cup at me in a salute while taking a seat.

"Okay, as Carl Sagan said, 'You have to know the past to understand the present,'" I said. "Let's venture back to your first visit.

"One of my most vivid memories was that of Gadsden, the homeowner who held the hurricane party—despite all responsible advice and an actual legal order to evacuate.

"The police began warning that during a storm they couldn't offer any assistance to anyone along the beach. They also told people to write their name and birth date and next of kin on their arm in case they passed on."

"As I remember," Sherl said. "Gadsden's home wasn't one of the high-end houses that had become the norm. I don't think his attitude was dissimilar to many others living along the shore."

"Gadsden taught math at the High School. He was well-liked by the students and had received several awards," I said. "He once told me that what he liked most about math was its absoluteness. "With math, there are right and wrong answers, a sense of accomplishment comes with correct solutions. It is predictable and understandable," he would tout. When decisions were needed or demanded, he could be counted on to offer direction. Gadsden was not one to be enamored by nuance. He was not one for jogging in place. For pondering.

"He didn't have a lot of money to spend on recovery, as others did. Most of the newer, richer property owners had the funds to recover after each storm, maintain their access to the beach, and preserve their valued ocean views. Many had deep political connections and had used them to get the government to pay for massive protective structures like sea walls and extensive beach nourishment projects. These often cost more than the structures they were intended to protect. During that time, those projects got funded. Many of the Hurricane Apollo survivors were difficult to work with. They were property owners but not really residents. They lived in Troy Beach for a few months, and had little investment or involvement in the community as a whole."

"My first exposure to Troy Beach was when I drove here to attend the workshops," said Sherl. "Although it was over a year after Hurricane Apollo, many homes on the shoreside of Front Street weren't repaired. Some beach home foundations still resembled huge tables as if arranged by a community of giants—just elevated first-floor platforms atop posts and piers."

"Interestingly, the supporting members you mention," I said, "were built to code. Strangely, not the upper floors. The Federal Emergency Management Agency requirements assured strong foundations, but they didn't care much about the upper-story living spaces. It's expensive to be cheap. It's expensive for others, too. The flying debris probably did more damage than the waves. The area was a mess with housing debris after the storm and as you saw, the area was cluttered with housing, commercial, and boat parts for years.

"The houses that were totally destroyed were mostly built to older, weaker building codes," I said.

"The sandy beach shoreline hadn't recovered when I was first introduced to the area," Sherl said. "It looked like it had moved permanently landward. The few dunes that survived still didn't have grass on them. You could see where waves had washed holes in the dunes directing surf and beach sand landward. These newly created corridors through the dunes directed the surf to homes and businesses that their responsible owners thought they had protected."

I nodded without interrupting Sherl's reflections. "One of the things humans don't like to acknowledge is that nature is always in process," Sherl continued. "The shore is always moving, and we want it to stay in one place so our immobile buildings and roads will be safe. It just isn't going to happen."

"Dead on," I commented. Sherl was on a roll.

"When we first visited the damage, I could see how waves had caused a patchwork of damage and repair. Beach sand had compromised roads, building foundations, and a host of infrastructure. There was still evidence that surf and high tides had washed onto Front Street. Landward of Front Street, you could see there had been heavy wind damage, but I saw little direct saltwater or wave damage.

Not until we got to the office and I saw pictures of the weakened foundations—the corroded rebar, rusting nails, and compromised structural supports—did I realize that the impact of these storm waves and salt spray extended far inland.

"Also, I didn't realize until I came here how bad the insurance situation was."

"Back then," I said, "very few of these houses were covered by wind damage insurance. Then, in the two thousand twenties, large companies stopped offering wind coverage completely. It was common, though not universal, for homeowners to have flood insurance. Without wind coverage, policyholders tried to get their flood policy to pay for all losses."

"Watson, not long after this introduction to Troy Beach, I discovered a large number of homeowners could have afforded insurance but didn't get it. They were well-heeled beachfront owners. They didn't need bank financing that would have triggered insurance, and they thought that they didn't need the insurance anyway."

"Many of the smaller homeowners who could not afford insurance had to sell to developers after the storm," I said. "The developers assembled the lots and built multi-unit condominiums. Even with new housing units built to stronger codes, more people were put at risk."

"Troy Beach had been in bad shape back then," said Sherl. "Many of those homes in the back of Front Street, I tracked after you left the city. Many initially survived the driving hurricane winds and sea spray, only to be determined a total loss as mold set in. Mold made a house uninhabitable and unsalvageable.

"Then there were the stores that were destroyed. Workers didn't have income for weeks or months. Some people had still been paying off repairs from a hurricane that hit five or so years before Hurricane Apollo.

"Troy Beach was beautiful because of the sea, but it was also dangerous because of the sea," Sherl concluded.

"That's why Cassandra held the workshops," I said. "She was hoping that with disasters so close together people would be willing to talk, listen, and objectively want to plan for what will come. While the disaster was still imprinted in their collective psyche. She hoped the workshops would result in an agreement on core values and build some trust along the way. Enable our communities to think of longer-term strategies."

"Sherl, that's what I want to chronicle. If Troy Beach could re-organize, others could benefit if they just knew what success looked like. They could create their own stories. Other places are just lost in political in-fighting or budget-busting legal battles. Cassandra worked some kind of magic here, hidden under her big smile and public process." I was privileged, being her intern and knowing what she believed and wanted for Troy Beach. At meetings, she seldom offered these insights directly. She would concur or subtly refute what others offered. Like a maestro conducting the philharmonic, she directed the participants without offering a single note. I am not even sure she realized she was doing this, but I think she felt that ideas coming from her didn't carry much weight. At the end of her meetings, the whole community believed they originated all the new directions for the future.

"The workshops were key," Sherl agreed. "But they needed the follow-up to result in anything. Residents are not design professionals, not professional planners. Your book needs to be more than the workshops. Maybe detail them in an appendix."

"I'll think about it. Be great if we could make a video. Spend a few months here on someone else's dime." I grinned and held up my coffee to toast Troy Beach. "It all started with the workshops."

"Back then, Cassandra had assigned me to take you on a tour through town," Watson said. "At the time I thought I was going to be a babysitter for some outsider who didn't know anything about our problems. I'm happy to admit I was wrong. She said I'd enjoy it. I hadn't worked with her long enough to know she's usually right."

"Yes, my friend, she called that one. But back to the past. Remember the marina? We met the Admiral, as you called him. Trying desperately to keep the operation together."

"Mr. Nguyen. From South Vietnam. I really don't know his rank with the South Vietnam Navy, but everyone called him the Admiral. He was our Harbor Master, and he was the main reason the marina existed. As you saw, it was long after the storm and there were still boat debris and a few pleasure craft still stranded in the parking lot. The fishermen were mostly Vietnamese and worked hard to repair their crafts and get back to fishing, their source of income. The Admiral described the marina as both a place and an organization."

"It's starting to look more like a functioning port," the Admiral had said, directing the statement to Sherl, with his strong accent. "Can't fish if you've got no boats. Fishing boats are strong. We have repaired or replaced most, and have started fishing again. But the fishery has not been doing well. Not many fish. We are still repairing our homes. Most of us live in Hilltop. There is no flood damage there, but there is a lot of wind. My roof blew off the kitchen, and I had no power for weeks. Thanks to Raj's Dirty Kitchen, we were able to eat. Families had a place to cook in Hilltop."

Raj ran a motel near Hilltop, and during disasters, he opened a communal kitchen in a large open area behind the motel's kitchen. This was referred to by all as the Dirty Kitchen. Mrs. Enicco, a local teacher, branded the area with the term remembered from her home back in the Philippines. Her family had an outdoor kitchen she referred to as the Dirty Kitchen and the name stuck. The large open cooking area included picnic tables, canopies, and gas stoves. Hilltop seldom flooded, but strong winds caused damage, including the loss of utilities and the community depended on this outdoor kitchen.

"The hurricane eroded the boat ramp and blew the boat crane over." The Admiral continued shaking his head expressing worry about the future. "Never happened like this before. Our repair shop was also washed onto the parking area, where my wife's little food truck is. Eunice, our local climate expert, says the seas are rising, and storms will only increase. We're a marina. We can't go anywhere. Who pays for repairs? And increased maintenance? Especially with the fishing being worse?"

"Perhaps the city will fix the docks and piers, ramp, and parking lot," I suggested at the time, not knowing the answer.

"For the crane and its repair, well that will be up to us, I guess. The protective rock seawall must be higher. We wrote a contingency plan after the last hurricane. Formed a co-op among commercial fishermen and bought group insurance. We're afraid we will not be able to renew the policy. At least for what we can pay."

"And then the Admiral summarized the problem perfectly," Sherl said. "I remember it well. He said, 'What's the future? Repair, only to repair again, and again…?' I'll never forget the combination of despair and dedication in his voice."

"Today, the fleet is now largely gone. But the port still provides a home to a small recreational boating community and even smaller fishing fleet." Sherl said.

"It still may help draw people to the city and provide nice revenue," I said.

"Back then, the fishermen were mostly Vietnamese, and their families resettled here after the Vietnam War. At the time, most had a lower income." Sherl said. "Their community has become more affluent. Many of their children are now professionals. Many have invested in the land behind Hilltop. They platted a new subdivision. Higher densities than single family. They should make a nice profit. You'll be able to see this when we tour the city. You should make some mention of these changes in your book."

"I'll try, but it may have to be just a footnote," I said. "I think the real story is about community migration."

"And, the societal changes that accompanied a changing climate," Sherl said.

"From the marina, we walked along the shore," I said. "Many smaller older beach cottages were in the process of being replaced with higher-end and higher-density residential buildings."

"Houses were being elevated higher and higher with each storm. All in compliance with the city's stronger development codes."

"Remember, we saw a few structures still standing. Had little damage. They were built to the city's higher codes adopted after earlier hurricanes. Rebuilt as recommended in the city's Hazard Mitigation Plan."

"At considerable additional cost, I might add," Sherl said.

"Not as much as you might think," I said. "Yes, retrofitting and repairing housing is always more expensive. Less expensive if done at the time of construction. More importantly, however, it's often not the best strategy along a changing shore. Beaches have eroded and retreated inland. This approach of requiring higher and higher first-floor elevations does not acknowledge that seas are rising, and the frequency and intensity of damaging storms are increasing. These standing buildings were made structurally stronger, being constructed to newer higher standards."

"I see where you're going," Sherl said. "Many of these structures became islands. They reflected the implementation of the city's Hazard Mitigation Plan, but ultimately all that resulted was that they became more difficult and expensive to demolish when the rising seas reclaimed the land."

"This concept of 'islanding' is receiving considerable attention in the academic community. As seas rise and waves wash farther inland, whole communities are becoming islands," I said.

"These storms helped the city rise up on the socio-economic escalator," I continued. "Not too sure if this is good. Storms changed the neighborhood. The post-event housing was always more expensive. If you couldn't afford new construction costs and costs resulting from implementing higher codes, you had to move. Sell your home to someone who can afford it. Many long-time residents and families left their shoreline locations and moved inland, which was a good decision for them, but the city lost good neighbors—social capital as academics like to call it."

"Did this make the traditional hazard mitigation and comprehensive planning processes counterproductive?" Sherl asked.

I nodded to emphasize my agreement. "New codes resulted in structures getting stronger and stronger after each event. And if the owners could not afford the costs, or when these stronger structures became islands, parcels got assembled and reassembled after each storm. Condominiums appeared as you mentioned earlier. Increased density means more people will be living in the higher hazard areas after the storms than before. You hit the proverbial nail on the head, and to me, it is the definition of being counterintuitive and counterproductive if we want to reduce the overall risk.

"But where Troy Beach received a few thousand dollars in tax revenue per lot before the storm, after the storm, and after lots have been merged, they were replaced with high-rise condos, and the city got many times this revenue from the same area. There's a perverse monetary incentive to increase risk."

"They always seemed to hunker back to adopting better building standards," Sherl said. It's what they knew. But these standards didn't reflect a changing shore. If replacing older structures with newer better-built ones worked, there would be no need to move."

"Continuing the memory lane tour, I remember us then walking farther up the beach to Gadsden's house, right on the beach," I said. "He was working alongside his friend, Ezra, the regional representative of the Volunteers Active in Disaster,

VOAD, organization at the time. Gadsden recognized me and began arguing as soon as we were in earshot."

"I am going to build my house higher this time," said Gadsden, as he stepped off a ladder being held by Ezra, "and make my home a little bigger in the process. Stronger! I am adding a lot more steel hurricane ties and nails than I did last time. All of them are hot-dipped and galvanized. Then I won't have to be told to evacuate every time there's a little wind and I can hold my hurricane parties in peace. And watch my neighbors run to their mothers' apron strings."

"This is someone interested in how you're recovering," I said, introducing you to the two of them.

"We're working as fast as we can," said Ezra. "People must have shelter. We've had many volunteers helping. Getting permits was a big problem though. That system can be streamlined but I feel we're all complying. We must thank all that have helped—the Mennonite Relief Organization, Mercy Corps, The Catholic Relief Organization, the Red Cross, the re-energized AmeriCorps plus all our local church groups. We want to get our area up and running as quickly as possible."

"Before the city changes the rules," Gadsden said in an artificially low voice, mockingly, but smiling as he spoke."

"There is a lot of talk around town about climate change," Ezra said. They're saying that the storms are caused by climate change. And that we should expect more of them. There have been many storms lately. More than I remember as a kid. But that's just the weather. Runs in cycles."

Ezra continued the discussion speaking more formally. "As for climate change, you know, the scientific community is divided. Watson, many scientists reject the idea of catastrophic human-induced global warming. There are sixty international experts in climate and related sciences who disagree. Extreme actions will hurt many people. Poor people mostly," concluded Ezra's delivery, given more as a memorized list of bullet points than a spontaneous response.

"I understand," I said. "You've been very vocal about this. But how does knowing more storms may be expected, regardless of the cause, impact what you're helping people accomplish here?"

"We are rebuilding our homes," said Gadsden. "If the storm comes again, we'll build 'em again and again, until we either get it right or can't afford to rebuild. We will not retreat."

"When will you not be able to rebuild?" Sherl said. "I am curious: When do you think you'll have to move? People have to make similar choices where I live."

"Money or circumstance," Gadsden answered as if talking to a younger, uninformed person. "I'll stay here as long as I can afford it. I was chatting with the city staff, and according to them, they are thinking of denying occupancy permits at some point. If my house becomes an island, I won't be able to afford repairs then, anyway. But that won't happen, so I have nothing to fear," he said while walking off to get some piece of equipment.

"You know," Ezra whispered, "Gadsden has no choice. He inherited the land and built the house himself some time ago. He can barely keep up with the taxes. For him, it is staying here and repairing his home with debris collected after each event or living in a culvert under Front Street. I am not joking."

Ezra nodded at his friend, who was in sight but now out of earshot. "Where do you think all the anger comes from? He is extremely vulnerable, and he knows it. There is nothing he can do about it. If he loses this house and the ability to rebuild, where would he live? Who will make room for him? What could he afford? America is not very good at building new affordable housing. There is a reason for his defiance. He sees himself as a victim. On many levels. Hard for me to watch, so I help out a little."

"Understood," I said. Gadsden returned with his tools. "Best of luck to you and we hope you never have to move."

We all shook hands. Gadsden and Ezra waved to us we walked away.

"That conversation with Gadsden and Ezra was difficult," Sherl said. "They were so determined but so misguided. I can understand why people are emotionally tied to their homes, but the emotion around climate change seemed odd to me. I may not be a scientist, but I am a science user."

"There were many people in the same straits," I said to Sherl. "Affordable housing was a huge issue then and is a huge issue now twenty-plus years later. That kept many people trapped in the area, trying to fix up and manage houses that they feared they were going to lose eventually anyway. They can't bear the thought of just walking away. Neither could I. Some would not survive bankruptcy being on their records; some had an emotional tie to a dream of visiting grandchildren, and some had no other financial asset than their house."

"You need to discuss in your book, what cities and the lower-income residents are up against. After every storm, it seemed, cities kept adopting stronger and stronger codes. Homes get higher, stronger, usually a little bigger, and more expensive after each storm. You pointed out that after Hurricane Apollo, a couple of home citadels survived. They became showcases."

"'Hope homes' I called them back then. I still do. Island homes in the making." I was torn between sympathy for the homeowners and frustration that all were caught in this cycle. And there was another indirect negative effect. With these expensive fortresses surviving the storm came the desire to protect them, to armor the shore, to spend more fiscal and political capital.

"When I was in the Peace Corps, in the Philippines, I visited communities of sea dwellers. The Bajou and Tausug. They live in houses on stilts, surrounded by water. On desolate reefs. Is this the future vision for Troy Beach? Floating homes might be an option if they were in a protected bay. Like Sausalito, California. What about amphibious homes that are on dry land until it floods and then float? That's a strategy being tried in the Netherlands."

"International Code Council building codes are the law of the land in many coastal communities in the U.S., like Troy Beach," Sherl said. "The codes do not promote residential 'construction over water'. And, the National Flood Insurance Program (NFIP) won't cover these homes anyway. Not now at least."

"Back then I was hoping to introduce you to a condo owner," I said while we were walking onto Front Street. "But we couldn't find them. The couple I knew well only lived in Troy Beach during the summer when they could enjoy the beach with their grandchildren. They wanted to relocate here permanently when they retired, but if they were like their condo neighbors, they would probably sell and move on with their lives once their grandchildren grow up."

"Ah, the absentee owner," said Sherl. "An interesting concern in many in beach towns. They often have disproportional political influence because of their affluence and privilege, but little interest in local politics."

"We had some ritzy mega homes back then, but Troy Beach was not a wealthy city overall," I said somewhat defensively.

"We went on to Front Street," I said changing the subject. "High water lines marked structures even after over a year since Apollo. The commercial buildings had been undermined and even collapsed in places. The sidewalks still needed repair. That surprised me. The street itself was clear of course, but we had to walk carefully through the downtown area."

"Where I come from, freshwater can wreak havoc during floods," Sherl said. "Saltwater is so many times worse."

"Sherl, the corrosive effects of seawater are cumulative. We didn't realize the effect of higher seas and the resulting saltwater intrusion. The impacts on the city's infrastructure were not evident until Apollo exposed them. This risk was underestimated in the Hazard Mitigation Plan. It is hard to determine the risks of what you can't see. We didn't realize the extent of the corrosion of our water and sewer pipes, and the resulting lack of maintenance certainly contributed to their failure. There was a lot of damage from the saltwater."

"The storm surges, winds, and driving sands took out windows, cracked walls, and ripped the contents into the street. In the downtown Historic District, fire followed the flooding due to broken gas lines. Mostly restaurants with natural gas stoves."

"I remember a couple of the buildings you said had been dining establishments," Sherl said. "The buildings were still burnt rubble. Restaurants tend to turn over ownership every few years anyway. Those were probably someone's dreams that were shattered in a single storm."

"I have to confess I can't remember what happened to every person that long ago," I said. "I know there were loans and help that came in. Most people made it through, but not everyone. Paige did, though."

"Paige had remarkable timing, buying a food truck just before the hurricane," Sherl said, "to go along with her long-standing Troy Beach Café. We saw her setting up the outside eating area in front of the food truck."

"Let's hurry," I said back then. "It looks like we've got a little time before she opens."

When I first met Paige, all I could think of was circles. Her head, body, legs—all circles. She wasn't fat, just looked like a colleague of warm, friendly and accepting circles. Nothing had changed, and circles, circles in a pink apron greeted me.

"Watson, welcome to my Café," Paige said while arranging her special homemade, real southern hot sauce on the picnic tables in front of what was left of her café. "Who's your friend?"

"Paige, this is Sherl. Cassandra asked me to give him a tour to show him how we're recovering from Hurricane Apollo."

"Sit down and have a drink. On me," Paige said while grabbing a coffee pot and cups. "I'm making more money now. After the hurricane. Go figure. Right afterward there were all the workers and the lookie-loos. I did well. More than during prior years. And, with my insurance claim, I should be fine. For now. Thanks to the food truck. The building is still a mess. It's unserviceable for a restaurant. Everything has to be white-glove clean if there's food involved. I was leasing the property, and the owner doesn't seem interested in doing much to it. My insurance money isn't enough to buy land and there isn't another site in town I want. So I'll stay with the help of my food truck.

"But the handwriting's on the wall. Each year we have higher and higher tides and threats from stronger and stronger storms. I could see you were coming from Gadsden's home and probably were also talking with Ezra before you came here. Ezra's a good guy. He helped me rebuild. But his belief in the coming end times doesn't help. Why put any energy toward a long-term problem if the world's going to end anyway is his lament."

"I guess a sense of shared external evil can bring folks together," I offered with a hint of hesitancy and a slight turn of my head. She caught the sarcasm.

"Kind of cowardly as I see it," Paige interjected. "What's frustrating is there are things we can do. I go to his church he does, but I seem to have an entirely different idea of how I will be judged. And, his friend, Gadsden, I feel sometimes, is just plain Looney Tunes. He will defend his right to rebuild, to his death. Probably, even after his house becomes an island or almost washes away. Fine. Let him keep rebuilding. It's his money. Mostly!

"Reminds me of my friend. She was told she had cancer. Early on there it was treatable, but she got a second opinion. Nothing wrong with that, but then a third and fourth. She finally found this guy from surfing the Web who was selling some herbal cure and had a massive following. I miss her."

"Gadsden is not bad guy," she continued after a reflective pause. "Really! He was the first place many would go to if they needed some tools or repair material. Also, he always looked pretty content. Satisfied with his life. His goals in life were attainable. They weren't grand—having a job, entertaining friends, being able to maintain his house. Until recently that is, when the world seemed to turn against him. I keep asking myself, what could he do to regain control, or at least the perception of control, of his life? At least he can rebuild his home as he wishes, and have his hurricane parties. He really got upset at the city when they tried telling him how to repair his home and to stop having those parties. Only control he feels he has left. I wish I could help more. I hate to see him so angry.

"Ezra's stopping a lot of forward-looking thinking. His words and actions are just not helping. His latest tack is telling us that we have lived a good life, asking us to remember the great times we have all shared and that we should feel blessed.

"Here's another cancer metaphor. He says that it is as if we were all dying of cancer or something. Not human-caused cancer, not cancer we can still cure, but that, that salvation is beyond our control. I ask you why would God cause our demise? Didn't he create us?

"On the subject of God. Ezra does pray for us. Not for our redemption so much, but more to wish for us to have the knowledge and strength to overcome adversity. And, he is the first to rush out with a crowbar and hammer."

"I can see how you two could have a conflict," Sherl said with no hint as to which side if either, he agreed with.

"I'm thinking about moving, but where? I grew up here. This is my place. What do you think I should do?" she asked, directing her question to Sherl.

I stepped in before Sherl could answer. "Beyond fixing your place, what do you think the city should do for the greater good? For all?"

"Stop climate change, protect the cafe, or move," said Paige. "One and a half of these strategies might work. Building a little dike could help our Historic District too. Could buy us some time. However, who would pay? Maybe it is best to try to find some other location." She looked at Sherl and me.

"I can't tell you what to do with your business or your life," Sherl said. "After all we just met. But somehow, I feel you'll figure it all out. You'll research the facts and determine a course that you're comfortable with. Live your life. That's the best I can offer," Sherl said while smiling.

"Paige smiled back. "Thanks. Sometimes it's hard to remember that part. Talk to Cap, the Bargain Beach Wear store owner up the street. He has some ideas."

"Many injuries occurred while residents were cleaning away the debris.

"Before you go, don't forget that a lot of injuries occurred while residents were cleaning away the debris. But not all were trials, tribulations, and trauma. We had some fun—well, at least it was fun when looking back. Picture me as a glob of grease.

"Now, as you have to remember, I cook Southern, which is another way of saying I have no fear of cooking oil. The storm took advantage of this fact, and the last surge busted open several five-gallon containers, baptizing my kitchen floor in slick, shimmering chaos. When I stepped in to clean, I quickly realized gravity was no longer my friend.

"Ezra, ever the problem-solver, sent two well-intentioned high school football players to help. Strong? Yes. Prepared for the physics-defying challenge of rescuing this roundish figure from an oil-slick floor?" she said, running her hands along her body. "Absolutely not."

"When they attempted to lead me across the room, my feet decided to ignore walking altogether, opting instead for something resembling speed skating. The real trouble began when I fell— a glorious, slow-motion descent into the pool of oil. My volunteers rushed in, determined to hoist me up, but every attempt had me rolling further. The boys kept grabbing at anything within reach, and I kept slipping away like a greased watermelon at a county fair.

"With their help, I finally stood, only for one of the poor chaps to realize exactly what body part he was latched onto. His response was immediate and dramatic—he didn't just let go, he *launched* me. I slid across the kitchen, gracefully colliding with the far wall. Since I was already here, I grabbed what I needed, kicked off the wall like some kind of sea creature, and swam toward the exit.

"Ezra had arrived to assess his volunteer, and discovered two flustered boys covered in defeat—and cooking oil. He assessed the situation and pretended he hadn't noticed a thing, lest he further humiliate his embarrassed volunteers. Ezra was not a man given to laughter, but you could clearly see him grinning as he looked away," she said while smiling, and directing her effort to arranging chairs. "I felt so sorry for those boys. Couldn't look me in the face," she said softly, shaking her head, as we walked away.

With the image of Paige tumbling around her café, forever seared in our minds, trying to temper our chuckling, we walked along the damaged buildings that made up the Troy Beach Historic District and up to the "Bargain Beach Wear" store. The building was standing, but the recently repaired sections still needed to be repainted. The display window was still boarded up.

"Wanna hat?" Cap asked as we walked up to a makeshift counter.

"Remember," I said to Sherl in the present, laughing. "He handed us that ridiculous Gilligan Island-type bucket hat. Which I have with me."

"As do I," said Sherl. Cap had explained that the hats were his way of documenting the hurricane.

"They are a little dirty," Cap said. "It was the saddest damn day of my life when I rushed in here after Apollo died down. I had to see what happened to my store. The inventory was strewn from here to halfway to Canada. Including these hats. But I kept 'em and they're dry now. I guarantee that they fit any occasion. You can still see our name on the front. And I still have the store that matches the name.

"I had nothing when I started this store. It grew with the city. Now I'm confused about what to do. I want to rebuild, but tourists haven't completely come back yet so the money's not here. If Eunice, is to be believed, our years at this location are numbered. Can't afford to keep on rebuilding. I used all my savings, along with insurance claims, to fix the place after Apollo, but flood insurance is getting more expensive, and I am a *small* business. I can see a time when insurance won't be available. This is my life. The building is pretty simple. Four walls and bathroom, but still, repair is costly."

"Do you have any longer-term plans?" I asked.

"Keep my investment. Keep the business going? I'm not wedded to selling this kind of stuff. But I am good at it. At business, I mean. I don't have a degree, even a high school diploma." He sighed and looked off into the distance. "Can't believe I admitted that. Don't share that often. It just came out. But I am just saying, I don't have a lot of choices. Who would hire me? I am good at what I've been doing these past years, and the community likes me. Maybe this city even needs me. Anyway, I have a lot of thinking to do. For the time being, I'll make repairs. I will probably build a storage attic and run the electrical cables along the walls, as high as possible to be safer from future flooding. There may be some other floodproofing measures I can take. The city passed out a FEMA booklet on the subject."

"Thanks. Good luck," we said in unison while walking away and wearing slightly soiled bucket hats with the words "Bargain Beach Wear Store, Troy Beach" boldly visible.

Visiting the mobile home park was next on our tour.

"You know when I first started interning," I said, "I thought mobile homes were actually 'mobile' and could just be hauled away when threatened. It didn't take long to find out that most are securely anchored. No towing tongues, no wheels, and numerous structural additions. Codes require mobile homes to be anchored with some kind of auger or tie-down. I knew this was intended to keep them from being blown or washed away, but many were anchored inadequately, and they washed off their supporting blocks anyway, but were connected well enough to make them difficult to relocate quickly. Somewhat of a lose-lose situation.

"Also, the park site is in a prime location, being close to the sandy beach. More expensive, much larger motor homes are replacing the older, smaller, more affordable trailers. It's just double trouble when there's a hurricane watch or warning."

Sherl agreed. "Mobile homes should have been referred to as stationary manufactured housing because that's what they were. Manufactured and really not all that mobile."

"We found Rover, the mobile homeowner at home and introduced ourselves." I said while returning to the benches we have found under a tree but still in the station.

"Great to see you. Just the man I want to see." Rover said to me as he pumped my arm in a long handshake. "I have some ideas I want to share. We were hammered by Apollo. Typically, only the more expensive mobile rigs got out. The rest of us just left our homes where they were. The city requires us to anchor our houses. They do not require the anchors to be detachable. To work both ways. Some people could disconnect the ties and move, but most got damaged. But even where moving was possible, many waited until the 'Hurricane Watch' notice, and by then they couldn't find a place to move to. The streets were packed. No safe places to pull over."

"We were just talking about how your mobile homes are not all that mobile," I told him.

"It's vital to understand that we are needed. This area needs workers and a lot of us live here. Outside of Hilltop and Valley, our park has the only really affordable housing in the area. My suggestion is to provide more mobile home areas and make it easier for the homes to be, well, mobile. Also, we need some safer refuge areas. That's about all I have to offer."

"He makes some good points," you said to me as we gestured a farewell and walked off.

"Unfortunately, there's a lot of negative bias toward mobile homes. Many of the city's residents viewed mobile homes as not bringing the 'right' kind of people or the right kind of structures. This rant is especially loud with our newer, more affluent beach residents. At the time I couldn't see the Troy Beach City Council allowing more mobile home parks as a solution."

"We could still add 'allowing for more mobile homes and providing more safe refuges' to our list of options," Sherl said.

Rover had lingered outside to tend to his small yard. Thatcher, Troy Beach's most prominent developer, approached us as we began to leave Rover's. When Rover saw Thatcher approach, he waved. "Bye, guys," he said and retreated to his house.

"Rover views Thatcher as a threat," I said. "She sees mobile home parks as a development opportunity, as I told you, Sherl. Thatcher worked in Troy Beach for years. She was responsible for most of the high-rises you saw along the beach."

"I see you are introducing a guest to our rising from the ashes like the Phoenix," said Thatcher while she walked up to us. "Though these are not now, MY condos of course. I built most but they are owned by the occupants. A few are rented long-term, some Airbnb, and a few are in a Timeshare program. At the moment I don't own any buildings. I have conditional building permits to construct two more along the shore, but work hasn't started. We're still in the design phase. A few townspeople have said I should reconsider the projects in light of Apollo, but really these storms don't hurt sales. Over the long run at least. Anything I build facing the beach will sell. And yes, I've bought lots from homeowners willing to sell and I consolidated them. I plead guilty to being a capitalist. To buy undervalued property and make it more valuable. Take me away." She held out her hands as if for handcuffs and laughed. "I paid a fair price and didn't twist any arms. This is the market system our country was built on."

"In the past, you created Limited Liability Companies, LLCs, built condos, and sold them," I said. "When all units were sold in a particular building you would dissolve the LLC and start all over again. You've been very forthright on this."

"I call it good business," said Thatcher. "The venture does produce a building. And I have not harmed investors. As for the buyers, there is a risk but 'buyer beware' is true for all real estate transactions. And don't forget that I provide housing, lots of tax revenue, tourist capital, and at some risk to me. There is always the chance that what I have penciled as being profitable. It may not pan out. Financing may fall through, interest rates increase, money becomes more expensive, labor problems arise, and you cannot forget those storms you are always mentioning. It is called sharing 'risk'."

"You may have just hinted at a possible solution," Sherl said, looking in my direction. "Inform buyers and sellers. Have government assume some of this risk for developers, as a bargaining chip. Public Private Partnering comes to mind, as does providing government support for enabling transferable development entitlements."

"That is worth thinking about," Thatcher said. "When confronted with known risks and with limited serviced land available, we developers need protection. We may be more willing to share profits. But please remember, I follow all the laws and leave it up to others to inform the public about whatever you want people to know. Climate change concerns have aired on the news for decades. If people don't believe in the science, it is not my problem. If people from up North, or anywhere else for that matter, don't take it seriously, why should I? I still can sell all I build. And, there's not anything wrong with that." She shrugged her shoulders and smiled. "See you later, fellows." Thatcher waved and walked away.

"After our discussion with Thatcher, we went to Raj's for lunch," I said.

"Ah, yes, the famous Dirty Kitchen. It is gone now. I can never forget how I could ever forget the Kitchen. The aromas, the clutter, the enjoyable and welcomed chaos." Sherl joked while taking a sip of coffee.

Building Your Resilient Story

Conflicts/Rising Action: Current and future threats, *assessed*. The challenge—confronting present realities

Part I presents aspirations, and Part II, this chapter, introduces obstacles and conflicts to be resolved in pursuit of aspirations. Referencing the WOOPs approach I mentioned at the end of Chap. 1, this process is called mental contrasting. Forcing the storyteller to view their wishes through a realistic, objective lens.

Accordingly, this chapter begins a crucial step in your storytelling process. Dr. Oettingen's research found that telling stories and formulating wishes reduces the storyteller's blood pressure. It has a measurable cathartic effect. This is good in and of itself. However, many of you want to effect change and to do this, there are subsequent steps. She found that if storytelling continues and the teller subjects her tale to discovery and analysis of realistic obstacles and a planning process, action is more likely. The teller will feel less complacent, but the approach provides sufficient motivation and knowledge to lead to action.[1] The story goes somewhere.

Action Rises:

Resilience is personal, and this aspect needs to be reflected in your character as you have them face almost insurmountable threats.

Here, you raise the action while our characters confront challenges, obstacles, or revelations.

Conflict results when aspirations confront realities.

Troy Beach conflicts and intrigue revolve around the difficulty our characters have in preserving their values, their known and accepted Normal in light of climate changes. (Fig. 3.1) After reading Chapter Three, you now know that Troy Beach is reliant on Hilltop and the Valley Subdivision and that any long-term resilient story will depend on each other. You also know more context concerning our story's conclusion. But you are only beginning to learn HOW they got there.

Plot Development/Conflict Resolution/Resilience Concepts

Your story and its resolution will be different. Let's examine how you may proceed with help from Troy Beach.

Who. In this chapter, Watson adds some characters—Thatcher, Rover.... He expands the roles of those you had been introduced to—Paige, Gadsden, and Ezra. A major theme of our story is that resilience is personal. Your characters, including yourself, will drive your story.

---

[1] Oettingen, Gabriele, "Positive Thinking Inside the New Science of Motivation," Penguin Random House, 2025.

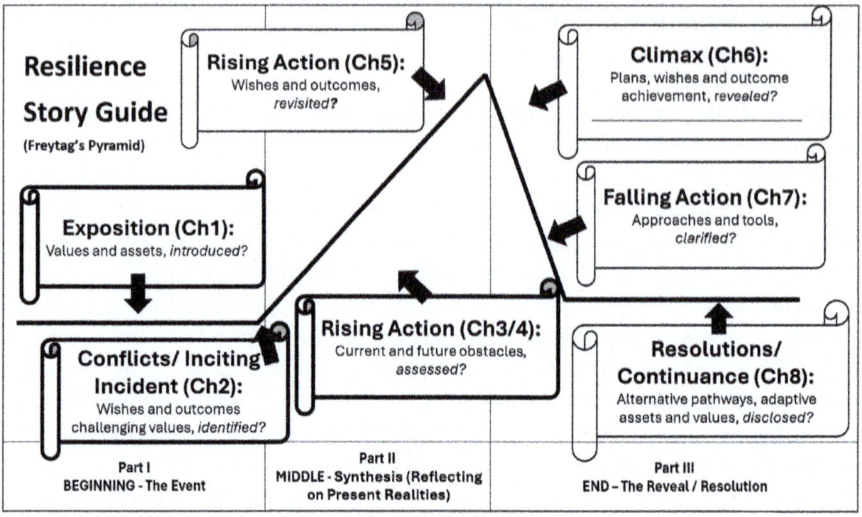

Fig. 3.1 Freytag's pyramid—Obstacles assessed. Prepared by Bob Freitag

What: More layers were added to the story context, mainly regarding WHAT changes had impacted the characters lives.

Where: More information is provided about the Valley Subdivision and Hilltop. These communities are gaining importance.

When: The Chapter continues 25 years in future. The dialogue, however, bounces back and forth between that remembered from the Apollo initial recovery phase, to the "present"—25 years in future. This recall was done to enable various characters to introduce and build upon their WHY and HOW musings. Your approach may be different. You may begin with a different phase in emergency management.

Why: This chapter builds on WHY damage occurred and that adverse climate change threats will abate any time soon. A few characters are beginning to consider HOW they will need to recover differently.

Although the story begins to project the need for re-organization and thinking of recovery differently, most of the discussion of HOW to prevent future damage remains anchored in more traditional hazard mitigation approaches. Approaches where vulnerable assets are either moved, accommodated, or protected. (MAP).[2,3]

Troy Beach's trajectory, as advanced in the Comprehensive Plan, is still viewed as the WHAT to which the communities should bounce back. This is what most residents consider normal, but this NORMAL is beginning to be questioned.

How: Although a few residents are starting to question the pursuit of existing plans, there appears to be a recognition that the current trajectory is unsustainable.

[2] Freitag, Bob. Floodplain Management: A new approach for a new era. Island Press, 2008.

[3] Doberstein, B., Fitzgibbons, J. & Mitchell, C. Protect, accommodate, retreat or avoid (PARA): Canadian community options for flood disaster risk reduction and flood resilience. *Nat Hazards* **98**, 31–50, (2019).

Acquiring the ability to reorganize is entering the dialogue, and resilient thinking is becoming more prominent.

As you continue to build your story, continually question your choice of resilience lens. Are the ones chosen appropriate? What are the obstacles? What's achievable? Epiphanies?

Explore WHY your character's values and supporting assets are threatened. Review how you profiled your change agent(s). Think thresholds.

Paige's Story

If you're using our Resilience Story Board Template, Fig. 3.2, offers an example of how Paige may have continued her story. She envisions a levee providing temporary protection and providing time to come up with a better solution. This is a merging of bounce back better/bounce back different approaches.

Remember, resilience is personal, your story is about its characters. They are why you are confronting your problem. Just as Watson is not writing a resilience story

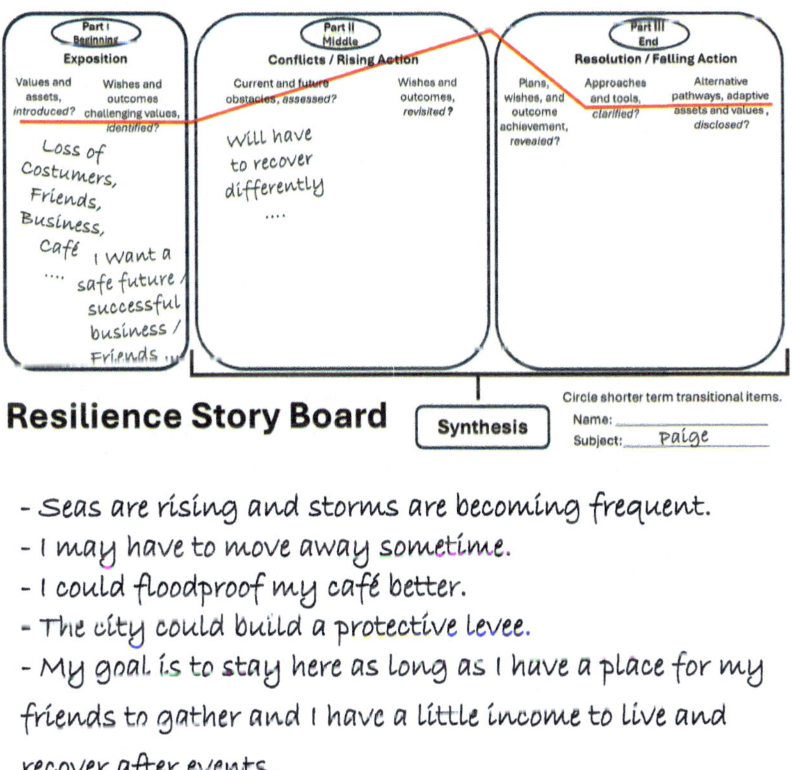

Fig. 3.2 Paige's conflicts assessed. Prepared by Bob Freitag

about an amorphous Troy Beach, he is writing about protecting the values of Paige, Cap, Ezra, Rover, and others you will meet as you continue.

Chapter 4
Sherl Visits Raj's

Current and future threats, assessed. The challenge—Contrasting present realities

Abstract The "Dirty Kitchen" was behind Raj's motel, and it, along with the motel and restaurant, had become the center of community life following Apollo. Survivors told and retold stories, and in the process created a new community vision. Conflict continues to build, but there is an opportunity to discuss, argue, and suggest resolutions. You are asked to clarify your conflicts and introduce possible alternative resolutions.

Keywords Reorganize · Climate change · Housing · Insurance · Permits · Workshops · Obstacles · Agora—dirty kitchen · King tides · Strategies

New Troy Beach Setting, Coffee Shop—Remembering the Survivors, Continued

The "Dirty Kitchen" was behind Raj's, and it, along with the motel and restaurant, had become the center of community life following Apollo.

Raj bought the motel a few years before I became an intern. After the new highway was completed, hardly anyone used the old route anymore, so few stayed at the old motel Hilltop's most prominent resident at the time, Ozell, tried to buy it, but his offer was rejected.

Raj had just emigrated here from India with his family and was looking for a business to purchase. The previous owners sold it to Raj because they didn't want to sell to someone from Hilltop.

B. Freitag, *Resilient Shores: Adapting to Climate Change Through Storytelling*,
https://doi.org/10.1007/978-3-032-09047-8_4

Raj had no idea that another offer had been rejected before he was accepted. When he connected the dots about the racial overtones underlying the sale, Raj made every attempt to create goodwill within the Hilltop Community. He hired residents including cooks and maintenance workers. The motel became known as Raj's, and the motel restaurant was renamed the "Chang'en Cajun."

To Ozell, who had tried to buy the motel, life was pretty much a matter of fact. He was Troy Beach's handyman, or at least that's what was printed on his truck. His standing contracts with the Historic District led to him fixing leaky pipes and adding electrical outlets, but he would also stand in as a cashier when needed. Everyone in Troy Beach and Hilltop knew him, liked him, respected him. Owning the motel would have been nice, but its loss did not linger. Ozell's wife came up with the name.

The old motel was very well built and with its use of glass and open courtyards, it represented a fine example of post-war 1950s modernism. Raj loved the style. The motel and restaurant were disaster-proof located above the level of flooding and located in a swale. Raj's old motel and restaurant survived hurricane-force winds pretty well. Rooms and the restaurant were typically open for business within a few hours of past storms.

Because of its location, the motel and restaurant didn't do well financially in normal years. The motel was full during the tourist season, but otherwise, rooms were not in high demand. Until there was a hurricane and FEMA dollars arrived. After the storm passed, Troy Beach residents and disaster workers stayed for months at a time. According to Raj, if it weren't for disasters, he would have closed it many years ago.

"Remember when you first saw Belle?" I said adjusting my seat. "She was sitting at one of the tables in the Dirty Kitchen."

Belle was right out of a Southern Gothic novel. Somehow just sitting there, you could tell she was the most important person in the room. Any room. Maybe it was her age or the way she wore her clothes—they were casual, but she still seemed formal somehow—but you knew she ruled the town.

"Sherl, I think she was the oldest resident of Troy Beach and definitely the most respected. Her family owned most of this land before the Civil War."

"Come join us. We have a lot of fun here. You'll hear hurricane war stories, and actually some may be true," "Belle." Said in her welcoming southern accent.

"I remember that first interchange well, Belle winking at us," Sherl said looking directly at me, assuring that I was listening and not studying my surroundings.

"Recipes are shared here, and there's always a lot of laughter." Belle continued. "Some of our culinary experiments here make it onto Raj's Chang'en Cajun menu. You must try the fried Mandu Samosa dish or Mrs. Oh's Reuben sandwich with a hint of kimchi. And then there's the legendary, at least around here, mango pie that I helped create with Mrs. Ala'ilima. It's sort of a key-lime pie, only made with mangos."

I told Sherl I thought Cassandra may have had something to do with creating the Dirty Kitchen. She and Raj were close. I'd never been bold enough to ask her.

Ozell joined us, as Belle continued talking. On most Fridays, he would drop by Belle's to mend a screen door or replace a deck board. When he was there, Belle and Ozell sometimes had tea and talked about the townsfolk.

"We're just about to go inside and order," Belle said. "I'm having the special. Black-eyed peas, mustard greens, and ham hocks spiced up with a little curry. Ozell will have the chicken adobo potpie," she said, speaking for Ozell. "Which he always has, I might add. The adobo pie is teacher Mrs. Eniceo's recipe although I helped with the crust. Just think bacon drippings."

"You know," Belle said while looking at us with added intensity. "Not all would be pleased with our experiments. I am confident that Grandmother Eniceo, along with other grandmothers, are waving their heaven-sent perturbed, disapproving fingers at their daughters. I am not sure that generations of passed-down research and resulting cultural perfection being traded for a little culinary humor would be appreciated. Kimchee Adobo or Pancit, the nerve. Ham hock curry. Still, the bravery expressed in these sectarian efforts was a lot of fun."

"We're high and dry, and on bedrock." Ozell said, politely waiting for Belle to complete a subject on which he had very little interest. "Well, not completely dry if you consider the rain," said Ozell. "Hurricane winds raise havoc, and the horizontal rain doesn't help. The streets flood, roofs are often ripped off, and we are the first ones in the region to lose power. However, we were still in better shape than Troy Beach."

"In fact, a lot of people are looking at the Hilltop Community as a safe haven," I said. "Then again there are a lot of armoring and hardening supporters. The 'let's rebuild stronger,' but without moving, folks."

"I'll make a suggestion," said Belle. "Build a dike, or at least put some armoring at the toe of the bluff in front of my house. The bluff's only six feet high or so but it's the highest land along the shore. That would certainly help. There's been a lot of erosion recently. My house has been in my family since before the War Between the States. It's a grand old house, once a thousand feet or so from the shore. Behind a Loblolly pine grove. Most of the trees have all washed out to sea now, and I guess the house is next if something isn't done. Our family doesn't have the money it once had, and a dike would be expensive. It would be great if the city could help."

"Lately there has been a lot of talk about Hilltop being 'up-zoned'," Ozell added. "Thatcher said we might make some money, but where would we go? Our kids are beginning to come back, and we are starting to become a real community again. My family has owned our property since Reconstruction even though we don't have a deed. No one to give us paper back then. Used to work for Belle's Great-Great, Grandfather."

"Worked for?" repeated Belle. "You could say that. We may even be related, but I haven't done any tests to prove it. I don't want to know, really. I just assume we're family and leave it at that."

"Funny that our Hilltop Community is becoming so popular," Ozell said softly as if he were reflecting more than communicating a thought. "None of the Troy Beach 'respectable set' would set foot in Hilltop for years. Now they're thinking of moving there. Can you believe it?

"We're having some discussions with Thatcher and the city staff,: Ozell continued.

"We're tossing around some ideas. I've included my son in the discussion along with the Admiral's My son is a land use attorney up north, but can practice here. The Admiral's son has experience with housing development.

"I really cannot see any advantage with being up-zoned. Or, being integrated with the city. But, I am not sure we have a choice.

"Also, of no minor importance," Ozell said. "Hilltop is a community. We have our own church, stores, the barbershop, just to mention a few activities. We evolved over the past hundred-and fifty plus years. All fit together. I can't see how anyone can design a city that can replace what we have.

"Just found out that Cassandra was holding a workshop. I may need to be a voice for Hilltop. I need to know how the city will incorporate what we value, into their overall plan."

A high school-age girl had been sitting near us. She got up and approached our table. "I have some ideas." Her voice was soft but filled with determination. "My name is Allison and I have as much right as everyone else to be involved in future of my town. I overheard you talking about the future of Troy Beach. We need to go green. People may not totally accept global warming, but most are aware that our climate is changing. Storms are increasing. Even Ezra acknowledged what was happening. That there is no end in sight."

"I remember you from a school assembly. You also mentioned greening up future development there," Ozell said. "Some of our children are returning. The Barber's son has come back from living up north and is building a new home in Hilltop that will rival anything you see on the beach. He does something with web marketing and the Barber said he was thinking of installing a solar roof. To save money and have reliable power. Did I mention that our power cuts out pretty much every time the wind blows?"

"The city should, like, require solar for all new homes. Hilltop could become stormproofed as it grows," said Allison.

"Wait a minute," said Ozell "As I said, we don't see any benefit from outside growth. We're fine the way we are. And with new housing comes higher taxes. Our wages have not kept up with the rest of the region. If we sell, where would we move to? There isn't anywhere close to move our community to, and let's be frank, we're African-Americans. We are not welcome everywhere. If there's a way, we could form a partnership where we could profit too, I'd be interested. Even some kind of joint ownership."

"We were wealthy, once, and owned much of the land around here, but no more," said Belle. "In a few years, my house may very well wash out to sea. Then what do I do? Sure, I can see Hilltop growing, having more houses, but in the process, becoming more expensive. Too expensive."

"I don't know much about that," said Allison. "But there must be a way to become free, like, of fossil fuels and have homes people can afford."

Sitting in the train station, I looked at Sherl and asked, "Remember the Russian-Ukraine War? Do you see Hilltop as Ukraine and Troy Beach, along with the developers, as Russian invaders, demanding their privilege?."

"You were a little over the top back then, Watson, but I understood that the only common ground to be that the Hilltop Residents felt the siege was inevitable. Unless it could be seen or developed as a partnership. Where benefits were shared."

"Remember Belle's neighborhood atop the eroding bluff," I added. "A couple of buildings slid onto the beach after the storm. Several more found themselves closer to the edge of the bluff and the buildings were not safe to live in. They were red-tagged and the city received a lot of pressure to allow these residents to reoccupy the houses. Some living in the area believed the city should reinforce the bluff. It would have taken millions. Just to protect a few homes. There were a few older plantation-style homes set far enough back to remain safe, but even these became increasingly vulnerable after the storm."

"Back then, during my initial exposure to the community, there was a lot of discussion about the Hilltop community," Sherl said. "Geologically the neighborhood rests on rock, a walkable distance from the city beach. A long walk, but still walkable. It's a couple of hundred feet above sea level. It was the result of some sort of land grant under Grant. Grant the President. And is the only high ground around."

"A Grant, grant?" I said, mulling over a possible wordplay to include in my book.

Many residents had issues with their property-titles. Families had lived there for generations without a title. Some Vietnamese immigrants moved to Hilltop. Lately, a few first-generation residents from Latin America and the Caribbean have taken up residence there. Hilltop was insulated in topography and socio-economics.

To give us a better read on Hilltop, Cassandra had set up an appointment for us to meet Angela.

Sherl smiled. "That was the first time I ever heard about her National Baptist Convention Church. I did know of the Valley Southern Baptist Church."

At first glance, physically, Angela seemed average. Then she smiled and the sparkle in her gaze projected an unguarded acceptance. I was welcomed. I was meeting her for the first time, I felt I had known her my entire life.

"We entered Angela's office and shook her hand," I said. "I recall that this was the last time I ever shook her hand. From then on, I was greeted with a hug."

"Got to enjoy those, still do," Sherl said.

Angela was in her late twenties but still conveyed confidence, demanding attention. Not in a threatening way, she just assumed to be taken seriously. She was tall, was dressed well without pretense, in respectful somber attire. As a youth, she was a popular girl who received top grades. The American Baptist College was very proud of her when she received her divinity degree, an unusual degree for a woman back then. By the time she ventured out into the larger considerably more intimidating world, the die was cast, she was unflappable and often amiably defiant. She could easily use her voice as a weapon, shield, or to affirm and embrace. In this sense, she was somewhat the opposite of Cassandra who worked best as a member of the chorus, while Angela did not shy away from center stage. Both had come to embrace and enjoy their roles. They worked well together.

At first meeting her, she greeted me with reserve not to make me feel uncomfortable, She warmed up quickly. "It looks like the church withstood the hurricane pretty well," I said then to Angela.

"Oh, yes, we do not reside next to the shore, and we're built on rock. We've replaced the roof enough times that we knew what to do after the last hurricane. This building is not just protected by God, you must understand, but by solid engineering and several generations of experienced town folk." She smiled.

"I'm taking Sherl around to show him the area and to have him talk to people about what you value about Hilltop in particular and Troy Beach in general. What we can do for long-term hurricane safety or protection against climate change or anything else about planning for our future," I said.

"I know of both areas and the interplay between them. Hilltop has always been the poor stepchild of Troy Beach. We're on the hill precisely because it's away from direct access to the water and the much-desired waterfront property. Even the fishermen live up here because they can't afford to live right on the coast. That's for the rich people." Her tone was calm, quiet, and maybe even a little ironic. "Don't get me wrong. Our community is here." She spread her hands upward to indicate the land around her. "If you offered us a switch, to move to the shore, we wouldn't do it. But when we first came here, that wasn't a choice."

She paused to let that settle in.

In that pause, they all understood what would have happened. Recently freed enslaved people who had few choices were allowed to live on land no one else wanted. The more privileged continued to own their coastal land and priority farmland surrounding the higher, rocky outpost. But Hilltop lasted almost two hundred years and counting.

"The land is not great for crops and their existence is marginal, but they make a living. There's still room for Hilltop to grow. Behind our community are farms that are owned and managed by Hilltop relatives."

"That's probably why no one has tried to buy it, but these lands provide a lot of space for expansion," said Angela.

"You've grown deep roots here," I said.

"There's no reason to avoid the topic here and I think the word you're trying to avoid is gentrification. It's not that we don't know some in Troy Beach are looking at our neighborhood as, well, they might think of it as a safety valve for people who live on the shoreline. We might think of it as people with more money who want to take over our beloved neighborhood. An unreceptive takeover," said Angela.

"That's an unfortunate term, we can agree I think," said Sherl. "Gentrification implies that a vulnerable community is being ousted by those with greater resources. Seeing value in the weaker community. Or you can think of it as two communities merging. You can think of it as the larger community investing in the lower-income community. But once you call it gentrification, you've drawn lines that are hard to erase. I hope Troy Beach is not thinking of a plan to upgrade what they see as a bad neighborhood into a 'better' one. Troy Beach is looking for a way to protect themselves. It may be seen entirely differently."

Angela smiled. "I can respect that. If that's the way we approach this, maybe we can figure out a way to save lives, maximize respect for my neighbors, and minimize disruption for both communities. Gentlemen, I warn you I'm going to fight for those principles, but I desperately want a solution that works for everyone. So, consider me both an ally and an adversary, depending on what's on the table. Hilltop had been growing, but then many of our children went away to get college degrees and pursue opportunities elsewhere. Although lately, with working from home opportunities, some are returning."

"What ideas do you have in mind?" I asked.

"To start with, we will not put up with being condescended to. We are equal partners in this. We expect to be involved in all meetings and spoken to with respect. Let's face it," she smiled, "we have the literal high ground here."

I nodded. "Can't argue with that."

"Second," Angela said, "the character of Hilltop cannot change. We want to keep our hometown – its accepting atmosphere."

"You've raised a mutual concern," I inserted. "Nothing will happen in the next few years anyway. Many in Troy Beach don't even see a need for change."

Angela nodded slowly. "I don't know you, but I know Cassandra and if she trusts you, I trust you. For now. I won't be at the first workshop. It conflicts with choir practice. As they say, that's my day job. But I hope to be there at later meetings."

The discussion continued and though no single direction was offered, Angela was intrigued by Sherl. She looked forward to working with him and her respect for Cassandra increased.

"That was a key meeting, Sherl. We gained a huge ally in Angela and that was when I saw how much of an asset you were going to be in Troy Beach's process. Seemed like there was trust."

"Angela seems to trust most, initially that is. The benefit of the doubt. And she and I worked well together. There aren't many people like her. She says what she means, doesn't hold a grudge, and follows through with her commitments. I can't wait for you two to meet up again."

"This seat is getting hard," I said. "I really want to run out that door and see the new city," I warned my old friend.

"Only one more place to review," Sherl said while lifting his coffee cup. "The Valley Subdivision. On your first tour, we didn't have time to visit there, but it is still important."

"The Valley River that flowed through the community had always flooded, and flooding followed most storms. Even before I became an intern, the flooding was becoming more common, and the flood elevation was getting higher. If one could put up with the inconvenience of having to make repairs and clean up the mess every couple of years, it was a bargain. Because they were river floods, their floods always receded. Not like Troy Beach where the shoreline was actually relocating inland, permanently. Because of the topography, the large lower flat river bench meeting steeper the valley wall, flooding was isolated to recognized areas of the floodplain. Flooding was just deeper and occurred more often as the frequency and intensity of the storms increased."

"Many people see the National Flood Insurance Program (NFIP) as an enabler," I said. "Enabling those living in unsafe locations to keep living in problematic areas. This is especially true in Valley. Here you could think of the NFIP as a housing affordability program. Within the Valley Subdivision, flooding enabled lower and moderate-income housing to exist."

"This needs emphasis in your book," Sherl said. "Having the NFIP enables affordable housing to continue."

"This was obviously not the intent of the program. What Ezra told us as we were leaving stuck with me. There is little affordable housing in the region. If we eliminate floodprone homes, the inventory of available housing will be reduced forever. If owners retrofit these homes, making them safer, they would likely increase in value. Removing the flooding risk and associated stigma also reduces the availability of affordable housing. We as a country cannot seem to build ***new*** affordable housing. Floodplain management may be as much about enabling the continuance of inexpensive housing as it is about managing the floodplain."

"Maybe. Seems like there should be a more direct way of providing affordable housing." I said.

"Now we have completed our Troy Beach refresher course," continued Sherl, "you should have a good idea of what Troy Beach used to look like. Are you ready to see what it looks like now? I think you will be pleased."

I leapt out of my chair, eager to see what happened in Troy Beach during the years since I last saw it.

We got up, walked out of the station, and found a historic marker illustrating a map of old Troy Beach. We stopped.

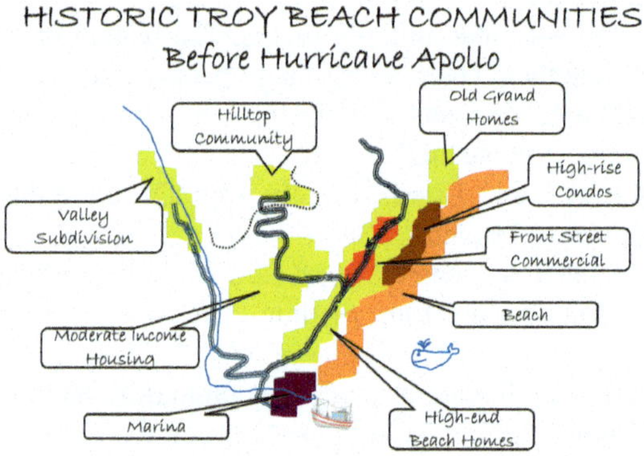

"You know, Sherl, Troy Breach was not unique as far as beach towns go. There's the sandy beach. The principal asset and from where all community values seem to emerge. Shoreward of the beach was a row of high-rise condominiums. They formed a real and psychological wall along the shore."

"These," said Sherl, gesturing to the brown-colored areas shown on the map referencing higher density areas, "were new at the time, and all were probably built to code. When we toured the area, most condos were fairly new and came through the hurricane relatively well. At least better than older structures along Front Street," Sherl said while pointing to Front Street.

"Retrofitting and updating older buildings had a history of working." Sherl sighed. "as did reconfiguring the rivers."

"Back home, my city spent millions of dollars straightening out river systems to make it easier for fish and water to flow through and realized it was a disaster. Then we discovered that the fish needed riffles, runs, and pools to thrive. We had to go back and spend more millions to put in boulders, anchor downed trees, and rebuild wetlands along rivers. Nature knew what it needed. It was cheaper for humans to adapt."

I nodded and pointed to the shoreland area illustrated on the map and said, "We bought some homes with FEMA Buyout funds along the shore. The good thing was that these vacated sites could no longer be developed with huge condos, but the area resembled a checkerboard of vacant land, older once summer cottages, sandwiched in between newer very high-end homes and high-rises. The city had problems servicing these scattered open lots. Rising seas and a storm surge destroyed or made permanent islands of any remaining development."

Moving my hand to Front Street, "Landward of these few blocks of development was Front Street and our Historic District."

"These commercial areas hosted beach-oriented stores and restaurants," Sherl said. "Many businesses suffered from the earlier hurricane, but not all. After Apollo, many bit the bullet, and repaired their structures to code. After you left, seas continued to rise, and high tides engulfed the properties. Destroyed infrastructure in the process."

"Landward of Front Street were our more affordable homes and where most of the year-round residents lived. And many still do. These lots were included in a FEMA-designated 100-year floodplain zone even back then. Now the area experiences extremely high tides monthly. We call them King Tides. You can see markers throughout the community (Fig. 4.1). The lands farther behind Front Street are in the FEMA 100 year floodplain, but these designations seem more a wish than reality. Even the homes farther back flood every few years."

"Out of town, up the Valley River, the Valley Subdivision floods every couple of years now."

"The residents of Hilltop were never threatened by flooding. Their threat was from wind and driving rain. And migrating populations."

"Yeah, as we discussed, Hilltop was, is, complicated," Sherl continued, pointing to the Hilltop area. "Like many older lower-income communities, the housing stock could benefit from some attention, residents needed jobs, and the servicing infrastructure could have been better maintained. There were many problems. There were tensions. Some people liked the idea of selling their houses and property at a large profit, so they were happy with where Troy Beach was headed. Then there were people whose families had lived in Hilltop for generations and liked the place pretty much as it had become and just wanted a little updating here and there. There were vacant homes and lower-cost lots. Also, workers were not as dependent on living near workplaces, and the larger Hilltop area had good internet service. All of this made the community more vulnerable to threatening migrations."

"Well, we'll find out more tonight at the Church. Let's go and explore the town," Sherl said as he finally stood up to leave the station.

Fig. 4.1 Information marker illustrating king tides as tipping points and thresholds. Picture taken by Bob Freitag

Building Your Own Resilience Story

Rising Action: Current and future threats, assessed. The challenge—Contrasting present realities.

This chapter introduces additional conflicts reflecting mental contrasting. The recurring question, given these obstacles, is "How is a community-wide resolution even possible?" Individual solutions, for Paige, Gadsden, Cap..., yes, but one that addresses the concerns of the entire community may seem improbable.

You now should have a better understanding of WHY climate change is critical, as perceived by our Troy Beach cast. You also probably conjured up theories of HOW the story may conclude.

Plot Development/Conflict Resolution

Rising Action

Following the exposition, as realizations, opportunities, and obstacles become apparent.

Questions

- What are the conflicts between your character's wishes and realities? Conflicts with where your characters are bouncing to?

- How do present realities impair/support your characters' envisioned strategies?
- Do any unanticipated opportunities/encumbrances emerge?

Who: Here, you gained a deeper understanding of the values defining Troy Beach, Hilltop, and Valley. WHY these values are being stressed, and HOW resolutions may emerge.

What: Climate concerns that can be almost impossible to address directly have surfaced through discussion among our characters. Fears of their homes being washed out to sea, a disappearing Troy Beach, the dissolution of businesses, and threats from gentrification have been raised. We hope that, in the abstract, your fears can be similarly confronted within a safe place—your story. An entirely new direction may be demanded.

As Watson tells our story, Belle would like the city to reduce a vulnerability—her eroding bluff—but she also questions the approach and is open to the need to relocate. Angela and Ozell understand their community is also threatened, if not by rising tides, by Troy Beach migrants themselves.

Where: As our story progresses, more intrigue surrounds these three communities—Troy Beach, Hilltop, and Valley. Resilience will differ for each community and for the characters involved.

When: Following the reflective dialogue introduced in previous chapters, the time bounces between 25 years in future and the immediate recovery phase following Apollo. It started a recovery story and is merging with one of post-event sustainability. Your story will undoubtedly differ.

Why: By the end of this chapter, you, should have a good understanding of WHY, both for individual characters and through their membership in the larger community.

How: Several HOWs have been introduced, and specific involvement strategies are gaining clarity.

Resilience Concepts

It is important to speak in terms of specific character-dependent strategies at this point in our story. Each of our characters is faced with devising how each may recover. For this, I will ask you to define strategies as approaches and tools required to implement an objective. This may help clarify HOW they plan to resolve their fears and become more resilient individually and as a community of individuals (Fig. 4.2).

Objectives suggest outcomes. Approaches offer the method to be used to achieve the objective, and tools are instruments needed to implement the approach. Gadsden elevating his house above the flood level would be an approach. A tool could be a zoning ordinance that requires new homes to be built above the level of flooding; however, for Gadsden, his tools are the actual tools he possesses (e.g., hammer, nails,

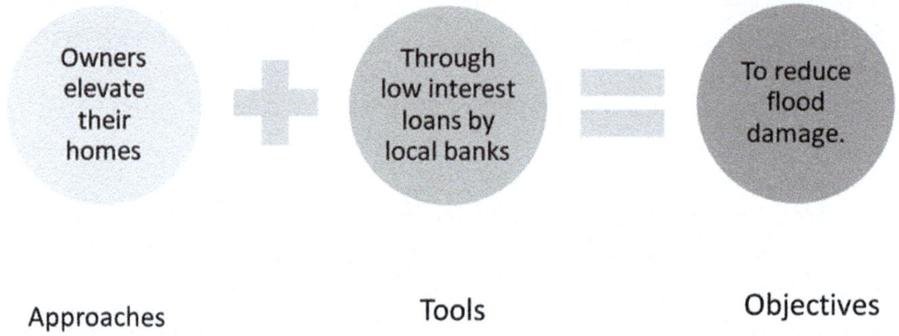

Fig. 4.2 Strategies defined as approaches and tools needed to achieve an objective

a saw) and the knowledge of how to use them.) An objective could be his value in having hurricane parties.

Paige's Story

Creating your story is more of a process of muddling through and sleuthing than anything linear. Even at this later stage in your writing, you are still probably introducing characters or character traits to help develop your plot. Similarly, your strategies are now taking form. Just make sure you end up making resilient the values you introduced in your earlier chapters. Complete your story arc.

Also, if you are having trouble presenting conflict, create a couple of characters, foils to bounce differing views off each other. Watson has Sherl as his discussion partner. Ozell has Angela, and Ezra has Gadsden and Allison.

Lastly, do not forget the importance of trust. When we began this process immediately following Hurricane Apollo, I was afraid that community factions would play off each other and create division. Different groups throughout our community would "Pick the target, freeze it, personalize it, and polarize it," as recommended by the noted community organizer Saul Alinsky.[1] Create divisions. Shore up power.

This didn't happen, largely because of Raj, Angela, and Belle. They were amazing. We often discussed issues well into the night in pursuit of common ground. The efforts of my staff should also not be minimized. They maintained a presence at the Dirty Kitchen throughout the planning process. Even Ezra was helpful, in his way; although he never accepted the concept of climate change, he accepted the threat and that impacts were real, and he never failed in assisting.

In your story, address the need for common ground and trust head-on.

Now, feel free to return to the Resilience Story Board Template and continue your entries. Paige's mental contrasting is confirming and expanding on the application of several resilience approaches. See Fig. 4.3 to see what Paige might write in the Chapter 4 column.

Continue telling and retelling your story to all who listen. Play around with changing your wish, obstacles, emphases, or with planning elements.

---

[1] Alinsky, Saul, Rules for Radicals: A Practical Primer for Realistic Radicals, 1971.

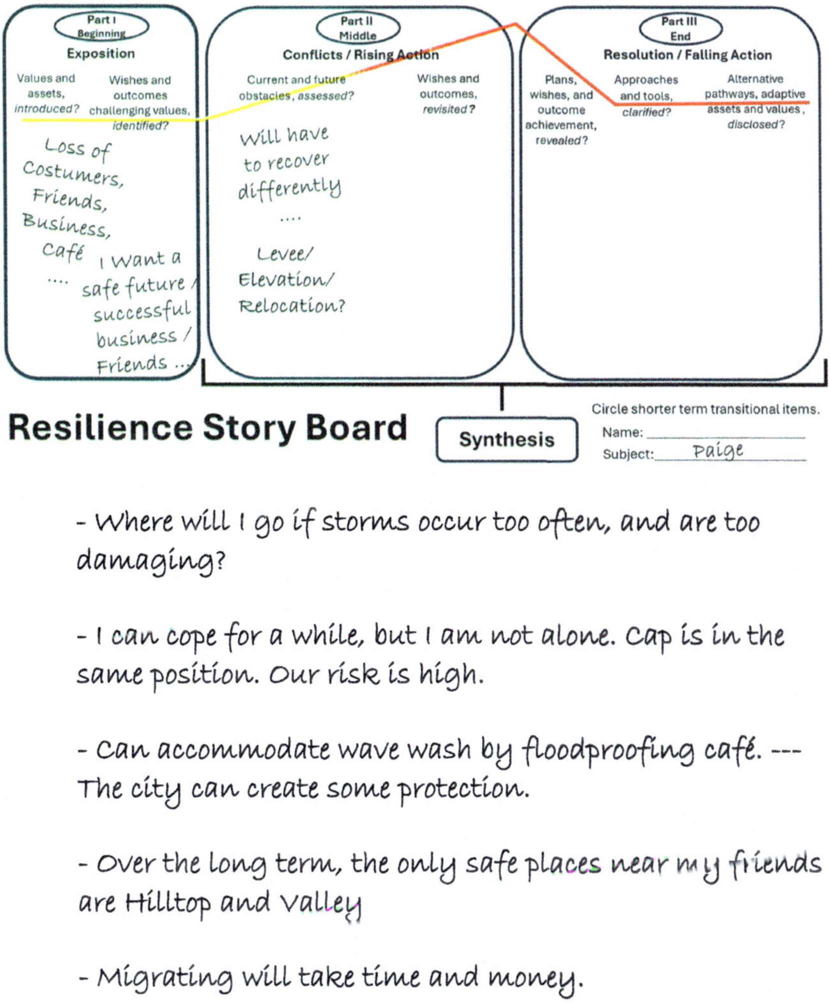

Fig. 4.3 Paige suggests possible resolutions. Prepared by Bob Freitag

Chapter 5
New Troy Breach

Wishes and outcomes, *revisited*. Challenge—Reassessing wishes. Clarifying aspirations

Abstract "Watson, you will see the fruits of your efforts today. You'll see what Troy Beach is becoming. Much of *where* you worked is gone, merged with the Hilltop community." Residents' visions are on full display in this chapter. A new, safer Troy Beach becomes visible, but HOW a unified path was determined remains a mystery. Reader/storytellers, here is where you revisit your wishes and make them more realistic.

Keywords Wishes · New Troy Beach · Valley subdivision · Hilltop · Historic District · Hazards mitigation · Climate mitigation · Adaptation

B. Freitag, *Resilient Shores: Adapting to Climate Change Through Storytelling*,
https://doi.org/10.1007/978-3-032-09047-8_5

NEW TROY BEACH SETTING, OLD OAK TREE—DISCOVERING A REVITALIZED TROY BEACH 25 AFTER HURRICANE APOLLO

Hilltop is the resolution

“Watson, you will see the fruits of your efforts today. You’ll see what Troy Beach is becoming. Much of *where* you worked is gone, merged with Hilltop”.

I eagerly walked outside with Sherl, not knowing what to expect, half dreading the thought of Hilltop having been torn down and replaced.

A smile spread across my face. I should have trusted my old friends. Of course, Cassandra, Angela, and the rest wouldn’t destroy Hilltop in the merger with Troy Beach. “Sherl, does it make sense if I say this looks like a ***new*** historic district? Some old buildings are here but in better shape than I remember. I don’t remember that bank, but it’s built to complement what’s been here long. So’s that. And that.” I pointed to several new buildings. New to me, at least. They might have been standing for fewer than ten years, but they honored the hundred-plus-year tradition of Hilltop.

“The city worked hard to keep the historic feel, their sense of place for this area. Look over here,” my friend said, pointing toward the coast.

We took a few steps through town, heading to our ultimate destination, the old oak tree that stood by the overlook. Hilltop had been a sleepy community when I worked in Troy Beach; now, I saw a city emerging. I could recognize this revitalized historic district, but the adjacent modern-looking commercial area was all new to me.

"I see a new emerging community. Mixed in with the older shops were more prominent high-rises along with a few construction cranes.

"Sherl, there's still a park-like feeling even with all this density. Plants are everywhere. Young trees are planted along the streets, promising to provide a future protective canopy. I see trees and shrubs on some roofs and on the upper-story balconies of the completed structures. Some plantings are in the open, while others are enclosed in what looks like clear glass greenhouse-like structures, seemingly staged to provide welcoming places to gather. Plants are even growing on the sides of some buildings. Pedestrian corridors and bike paths are winding around and through shaded public and private spaces. Few streets appear to be dedicated to cars".

"As was raised at the workshops, back when you were an intern, this is becoming a green city, Watson. Green does not only refer to being carbon friendly, you realize. But green is the color. The threats that drove your workshops were sea level rise and storm surge. Troy Beach is addressing that by migrating to higher ground. The current threat to this emerging city is now wind and heat. The wind threat is being addressed with stronger buildings and positioning buildings to deflect faster and more frequent winds. At the same time, the overall design also routes cooling breezes through the city. Plants and tree canopies will help keep the city's public spaces cooler and provide shade along the corridors you describe. I am sure you feel the light cooling mist spraying along these walking paths."

"You'll be happy to know that when I stepped off the train, the first word that came to mind was 'green'. It's exactly what the train station screamed at me. And the milder temperature was welcoming."

"Excellent. Mission accomplished. Or at least mission in progress."

We kept walking through town. People smiled and nodded at us.

"I recognize the old barbershop and the adjacent beauty parlor," I said. "And the old hardware store and restaurant. They look better maintained than I remember. As does the old boarding house."

"Those original structures are still in use. Many shops are owned and operated by the same families you may remember, and the boarding house now functions as a community center. It's called the 'Belle Center' and is named after our old friend. What you're seeing is the revitalized, as you said, 'new historic district.' The downtown core of the new Troy Beach."

"Is this all that's left of the old Hilltop community?" I asked, hoping for a negative response. "A few well-maintained structures are still in use? If they're still around, I want to discuss this metamorphosis with Angela and Ozell." Both were very protective of Hilltop. I had a feeling that this wasn't everything they had in mind.

"The residents of the old Hilltop have been heavily engaged in what you see," Sherl said. "The city created a Community Redevelopment Corporation (a CRC) to oversee displaced Troy Beach residents' migration and accommodate growth. Hilltop residents occupy most of the CRC's board positions. Property owners are also becoming development partners and sharing the profit. You'll most likely learn more about this at the formal celebration tonight. It's being held at Angela's church."

"Remember Paige and Cap?" he went on. "You can see their businesses along Main Street." He pointed out Paige's Café and Cap's store.

I turned and looked down the street. The two businesses were clearly visible and located at the end of the historic district.

"Both have stores here in town, but they also have mobile structures that operate all along the waterfront," Sherl continued.

"Paige has been active in community affairs, and her café is a gathering place for many Troy Beach organizations," Sherl said. "She even has a back room where locals can gather during tourist season when the main café is occupied. It's a great pride to her."

"Cap's store used to be 'Bargain Beach Wear,' and he relied on beach access and tourists. I see his new store is called 'Beach-sakes.' It's a play on words, like Keepsakes, but with Beach. Anyway, his inventory is broader now, and he is doing well with mobile and permanent locations. You'll find out tonight. I am confident both Paige and Cap will be attending the celebration."

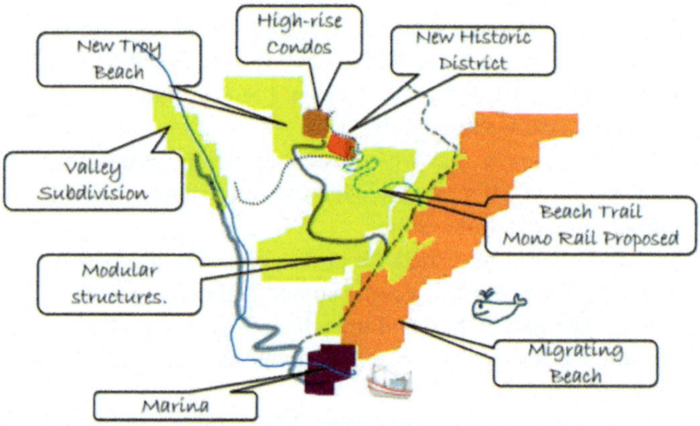

Troy Beach communities in adapting create a new normal. Prepared by Bob Freitag

We reached our destination—a wall alongside the oak tree overlooking what had become the older and now primarily abandoned Troy Beach I remembered. Our perch gave us an overview of the lower coastal shore where the old city sat. In front of the overlook was an information marker illustrating the future vision for the evolving community. Sea breezes blew up the rise onto our park, bringing the salt air and its scent of seaweed mingled with the minerals of sea salt and warm sand I had grown so fond of during my internship days.

"The new Troy Beach respects her old friends," Sherl said, gesturing to the massive trunk of the large White Oak tree we were standing under. "Look down. What do you see?"

I was astonished at the unexpected changes laid out before me. "I see what remains of the Troy Beach beachfront community," I said glancing back and forth between the marker, the new Troy Beach community, and the view of the shore. The shore landscape shown on the marker differed significantly from my memory. "The sandy beach is much larger than I remember. The shore's march landward was unopposed. Gone are the condos, most beach cottages, roads, and protective barriers. I see the remnants of a few home foundations, lots, and road segments documenting an apparent struggle. Signs of defiance. Of attempts by residents to stand their ground. Of the changing seas not caring and just bearing down relentlessly."

Sherl asked. "Do you see the restaurants, kiosks, dressing rooms, and restrooms along the coast? They're mobile. Made of modular elements that can be assembled and disassembled quickly should there be a hurricane and easily stored during the off-season. The beach locations are actually where Cap and Paige make most of their profit."

"I wouldn't have guessed that," I said. "That means tourism is back in a big way."

"Some parking surfaces covering the softer sands can also be removed. Do you remember seeing those metal mats they used in World War Two movies? The real ones are called Marston Mats. The design was adapted for use along the beach. They can be easily removed and stacked."

"The beach has grown landward," I commented. "More dunes are evident than I remember, and they're larger. Some marshland is established, and I see the start of a small emerging coastal forest."

It was astonishing to me how fast nature was reclaiming the waterfront. Whether animals would re-establish themselves was another matter. With the temperature of the water increasing, would this process continue?

"This reclamation needs a little assistance," Sherl said. "The climate is changing rapidly and the city has engaged nature scientists from State U to help with the transition."

"I also see what looks like newly developed areas. And a trail leading from this park to the beach."

"Do you see the construction along the trail? The city is building a monorail. At the series of workshops you were involved in, the shore and beach activities were identified as being key to the survival of Troy Beach. The beach was the reason Troy Beach existed in the first place and was its highest value. This public transit element is designed to maintain this beach connection. The beach will remain the driving value even with increased distance."

"The big question is, would such a trail bridge the physical separation between the town and enjoying the beach? Enabling the restoration of healthier and more expansive beaches should help," Watson commented.

"The development you see is set back from the beach, in the back of what was the old Troy Beach commercial area," Sherl explained.

"Yes," I agreed dryly. "I helped start the plan for its placement. Back then, the property was at the edge of the FEMA floodplain."

"It was determined by the city that at the projected rate of sea level rise," Sherl said, "this area should have a twenty-year usable life. This estimate established the land's value.

"The city gets some valuable tax revenue based on this value and use of the property. This helps pay for some of the improvements you see around you. The city purchased much of the property behind Front Street and leased back to developers. The homeowners' associations, in turn, buy these leases."

"The development you see are multi-story modular structures designed to be enjoyed, then disassembled and moved when warranted. If and when—I am confident you will be reminded of this phrase at the celebration—rising seas threaten the development, the development will be taken apart and physically moved to a safer location. Could be within the larger Troy Beach community, but other cities are beginning to build the infrastructure to receive these modules."

"I could never forget the phrase 'if and when.'" I laughed. "First, there were the workshops, then the follow-up."

The workshops were Cassandra's brainchild. Within two years into the first stages of recovery, residents were busily repairing their homes no differently than in years past. They were forgetting how hurricanes (or floods or any other disaster) came back time after time.

Cassandra knew she could scream her concerns as loud as possible, and no one would heed the advice. She needed to find a way to have the community themselves find common ground, reach agreement on the threat, and achieve agreement to some acceptable longer-term approaches. Just rebuilding stronger wasn't a workable reaction to climate change.

So, she decided to hold a series of long-term recovery workshops. All geared to build trust. Each would last a day or less, held a week apart to enable the community to mull over what was discussed and structured to include a lot of debate. She had experience with the World Café process and decided to use that as a starting point. This process enabled participants to refine issues through small group discussions. Clusters of contributors sitting around tables in an informal café type setting would discuss a topic. After some agreed period of time, these groups would rotate. One or more table group member(s) would stay behind to present their group's findings. Rotating teams would layer their thoughts onto those of the receiving table group. Cassandra used the approach to refine community values and contributing assets, determine threats to these values and suggest possible ways to resolve threat impacts.

The first workshop assessed common values and identified the assets responsible for their continuance. The second workshop identified threats to these assets. At the last workshop, participants devised as many solutions as possible, in other words, as many ways to reduce the risks to their values as possible.

She didn't necessarily expect agreement but a range of acceptable ideas. Local staff members would participate in the workshops. To help develop the approach and suggest implementation tools, she hired an outside consultant. Coming up with acceptable pathways she believed lay people could do, but developing a plan to implement demanded professionals. The Consultant's first and last names began with "L", and in college, he acquired the name L-Squared, or L^2, and it stuck.

"We tried to involve tourists as stakeholders," I said. "But could not get them to participate. I guess attending workshops was not part of their vacation plans. Cap was active with our local Chamber of Commerce, as was Paige. Both were dependent on tourists, so I feel that our visitors' values were not totally unrepresented. However, Cassandra was concerned with their lack of direct participation. She expressed worry that being represented by Chamber members forced us to view tourists as commodities and not community members. But we had little choice. This may be seen as a shortcoming of the process. However, in support of the process, we were not doing a tourist marketing study, but a plan for our more active local participants."

The first workshop identified shore recreation as the key value, with a healthy, accessible beach being the asset. This was really no surprise since Troy Beach only existed because of the beach—it was right in the name. Threats to the beach came from rising sea levels, increased storms and storm surges, and encroaching beachfront development. Many resolutions were presented and discussed. Migrating to higher, safer ground was discussed and not rejected, to be acted upon if and when the situation became dire.

It became dire.

Sherl changed the subject to how new development was occurring within the old Troy Beach.

"The developer, Thatcher" continued Sherl, "and Atlas, you may remember Atlas as Cap's contractor, along with the Ozell's son, are building these module apartment elements off-site and installing them, stacking them really, within current safer coastal locations. This isn't new. The approach was used way back in the 1970s in Japan and other places. Think of the Nakagin Capsule Towers."

"The concept was tossed around at our workshops," I said. "Construction with giant Lego blocks, plug and play housing."

"Even the emerging risk areas within some flood hazard zones have value," Sherl said. "With rising seas and increases in storm intensity and frequency, such areas were determined to have value and years of life before taken by rising seas. Some lands are dry now and can be used to return a nice income to the city."

"Until, of course, they're not." I frowned, offering an obvious question. "Are such transitory structures accepted? Can they get financing?"

"Our local bank has helped a lot," Sherl said. "Loans are now available. These leased-back developments are proving to be real money-makers for the city. They exploit prime locations and don't cost much to develop. They also don't represent a liability from a migrating shoreline. These units are factory-made, and we've cut construction costs. And they can easily be disassembled and relocated."

What about unions?" I ask. "When I was an intern, the building trades feared the lower wages that may accompany factory work."

"Atlas pays union wages at the plant. Also, this lease and lease-back arrangement has helped provide affordable housing. The construction costs are lower through increased efficiency. And because they can be moved, units can be resold. City revenue from the leases can help support affordable inland housing."

"If you look carefully to the shore," Sherl continued, "you will see more traditional mobile home parks and mobile food trucks intermingled with Paige and Cap's modular kiosks."

While I was working for the city, mobile home parks were almost universally renounced. What changed? Back then, only Rover and other mobile homeowners supported an expansion of the mobile home community.

"Necessity is the mother of invention," said Sherl. "Or, in this case, the mother of acceptance. They're part of the new need for mobile, modular structures. Troy Beach still benefits from tourism; some of these mobile structures provide short-term housing. The city is planning an amusement park on the shore. Also, it should be modular so the structures can be removed when threatened."

"Wow," I shook my head. "I wouldn't have believed so much could change in such a short time."

"When you're looking at sink or swim, survivors learn how to swim," Sherl answered. "There were leaders here, quiet leaders, throughout the community."

"That was another benefit of the workshops. Just getting people involved. Talking to each other. Who knew?" I said.

"Cassandra," we said together, looking at each other.

"The marina is still operational," Sherl went on, pointing beyond where we could see. "Located where you remember. There is really no other place. It's constructed at the mouth of Valley River, the only natural harbor for miles. The fishery has largely collapsed, so the fleet is small. In addition, recreational boats are becoming increasingly expensive, and the community of large, non-trailerable boats has diminished. The harbor does have a usable ramp and floating docks that accommodate several tourist excursion boats. These structures are protected from most storms by a system of higher protective revetments. The docks have storm anchors. There aren't any permanent structures in the marina. The Admiral, the old Harbor Master, died, and his son has taken over. Probably more out of respect for his father than for income. Future management will probably fall to the city.

"As for the Valley Subdivision, not much has changed. It floods more often, but still provides much of the region's affordable housing, but I don't know much more about it.

"Oh, I should add, State has introduced beavers to the upper watershed. They, the beavers, are beginning to have an effect. A lot of water is being retained. Flooding is still occurring downstream in Valley, but less than what was projected."

"What about utilities: stormwater, sewage, water, and power?" I asked.

"Troy Beach has retained its own public city-run utility company," Sherl said. "Along the shore, waste is composted where possible or collected. The city has an underground system to process brown and some gray sewage. However, gray water is largely recycled onsite. Storm water is also detained on or near sites, managed by an assigned host. These sites include low-impact development swales and detention basins. Water is made available through underground pipes in town. Nothing remarkable there. Cisterns and wells supply the lease-back developments along the shore. Waste is largely composted on-site. The module removable kiosks supply their own water.

"As for power, the city no longer uses natural gas or diesel. Everything is electric, and it comes from renewable resources. Most module and mobile structures along the beach generate their own power on-site, although the city has mobile solar-powered charging stations that can be taken where needed."

"That would make Allison happy." I smiled, thinking of the high school student who was so aggressively interested in green power.

"Look offshore." Sherl pointed. "See our energy farms? As you might have guessed, Allison is no longer a student. She and the VOAD Representative, Ezra, collectively manage our offshore wind farm and the solar fields that you can't see from here, behind the town. Allison is our City Resilience Officer and Ezra is the Solar Facilities Manager. The city has a grant with State University to build an experimental tide/wave-powered generator, but it isn't operational yet."

"So many changes. I'm almost speechless. I knew Allison was the valedictorian, but that was the last I heard of her. As bright as she was, she seemed a thorn in the side of Ezra. Allison was what we might have called a climate change influencer back then.

"Ezra was active in the community," Sherl said, "and organized our post-hurricane response efforts, but he had accepted his church's denial of climate science. Interestingly, she's rejoined the church.

"Not too sure if Ezra, has come to believe in climate change science. He may see the value of his job as providing low-cost power, not in reducing the use of fossil fuels.

"Energy storage was a problem, but as an element of our overall community energy approach, buildings and small clusters of buildings were designed as energy microgrids. They produce and now store power, mostly on-site. The city had planned to store energy in a reservoir. Pumping water up during periods of excess energy and releasing it when needed. A new battery has been developed that is sufficient to service these microgrids.

"Also, if you look above the marina, you will see the construction of the carbon capture plant. As a result of the major fossil fuel companies losing the lawsuit several years back, where they were held responsible for deterring the acceptance of global warming, the corporations had to build these recapture facilities. You have probably read that the plants actually make money for their companies. Seems that the lawsuit didn't say anything about selling manufactured products. The plant converts carbon into construction lumber."

"Go figure," I said. "Their penalty is making money."

"Well, that's your introduction to today's Troy Beach. Let's go to Raj's for lunch."

With all the change I was comforted by that the old oak tree remained, providing an appreciated historical anchor. It marked the heart of the city's new central park. According to Sherl, the tree would provide a welcome shelter from which to monitor and enjoy an evolving city within the context of the reclamation of old Troy Beach.

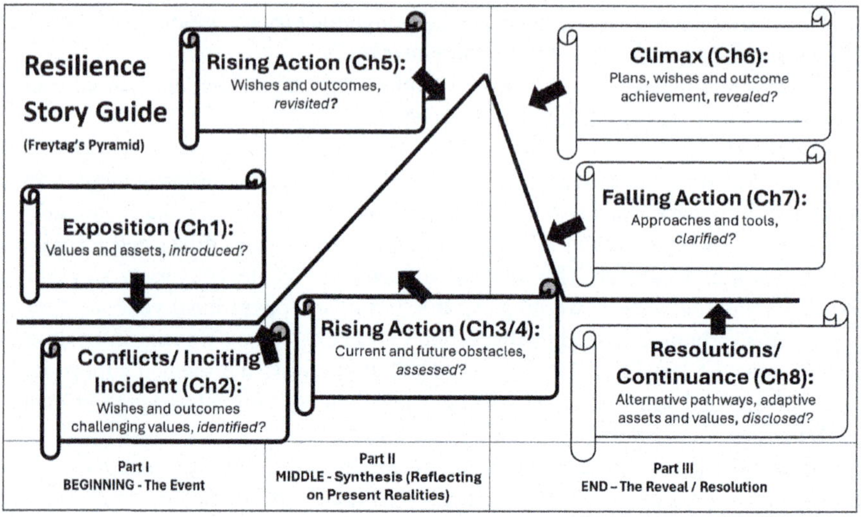

Fig. 5.1 Freytag's pyramid—wishes realized. Prepared by Bob Freitag

Building Your Own Resilience Story

Rising Action: Wishes and outcomes, *revisited*. Challenge—Reassessing wishes. Clarifying aspirations

This Chapter provides aspirational details glossed over in Chap. 2. We get a clearer picture of what Troy Beach is becoming—25 years in the future. But HOW was it achieved? You should be on "pins and needles" at this point. With all of this conflict between our characters aspirations being hampered by a changing climate. How did they arrive at common ground? (Fig. 5.1).

In our WOOP adventure, the chapter demonstrates that the process is reiterative. Think of looping through your story plot, adjusting as you refine your story. Telling your story to those who are willing to listen—many times. Redefining outcomes, identifying additional obstacles, and refining your wish, your aspirations.

Plot development/Conflict resolution/Resilience concepts

If you are writing a "who done *it*," or rather a "how *it* can be addressed so 'it' doesn't happen again," you are almost there. You know WHAT our three communities are becoming. You should understand WHY a change was needed but still do not know HOW the communities of Troy Beach, Hilltop, and Valley got there.

Who? Some characters' assets have changed. Their values do not. Paige and Cap have new stores. Some assets have not changed. The old oak tree is standing and Angela is still involved with the same church. Values introduced in early chapters have not been forgotten. The story arc has not generated a dead end.

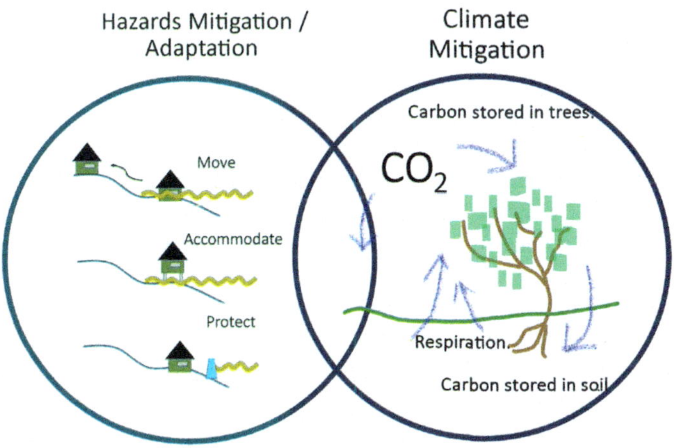

Fig. 5.2 The term mitigation can refer to strategies that minimize adverse impacts to hazards, as in "Hazard Mitigation" or to "Climate Mitigation," the capture of carbon. Prepared by Bob Freitag

You may have questions about how Angela and Ozell feel about the change. Do they see a protective, supportive new Troy Beach? We'll learn more about this in subsequent chapters.

By now you have probably realized that Raj is probably our most critical character. His making his motel available to the community allowed our public process. Watson didn't give him any dialogue.

Make sure your characters continue to drive the plot and that the values introduced are not forgotten.

What?[1,2] A changing climate was the root cause pushing the story, but it also demanded trust, agreement on common values, and recognition of cultural influences.

You also see how the new Troy Beach is addressing climate and hazards mitigation. Climate mitigation and the sequestering of greenhouse gases is occurring across all sections of the new communities. (Fig. 5.2) Ezra is managing a carbon sequestration plant. Allison is searching for ways solar, wind, and wave energy sources can be exploited. Even Atlas has discovered how he can help store greenhouse gases and render a profit at the same time.

As for hazards mitigation, the relocation of vulnerabilities is a crucial for this adaptation strategy, and accordingly, the city encourages mobile structures and even mobile infrastructure. They are tackling heat hazards by greening what they can and designing mechanical cooling systems such as generating mist along their walk paths.

[1] New Jersey, Department of Environmental Protection. Greenhouse Gas Emissions, https://dep.nj.gov/ghg/.

[2] Harvey, Orvis & Rissman. Designing Climate Solutions: A Policy Guide for Low-Carbon Energy. Island Press, 2018.

There are approaches that reduce carbon and mitigate or reduce a hazards impact. The reintroduction of beavers into the Valley watershed created wetland areas. These sequester carbon (climate mitigation) and reduce flooding downstream (hazards mitigation).

When? Our story began with the event and response, which led to an initial recovery and longer-term recovery. Here, Watson witnesses the culmination of the earlier stages of recovery he helped spawn 25 years ago.

Does your story reflect similar fits and starts, frustrations, and activities that evolve as your story progresses?

Where? Our story begins in Troy Beach. Then interest in Hilltop gains importance. Valley presents its own side story that is yet to be resolved.

You may discover alternative locations for various activities as your story progresses.

Remember, as detailed your starting story outline may have been, and whether it involved strengthening a vulnerability, bouncing back to some pre-event pathway, or reorganizing into a new trajectory, your plot, the story arc, will undoubtedly change as you, the storyteller, interacts with your characters, their values, and the change agent.

Why? Watson does not present a lot of climate science. However, through his characters, he does introduce climate impacts and why they are causing hardship. Why are his characters' values being threatened?

Resilient Shores assumes you have an introductory understanding of climate change science. Make sure when developing your story, you acquire a basic knowledge of the hazard(s) you are addressing.

How? How the threats to our communities' values can be minimized is the point of our story. This chapter offers some inferences, including trust, determining mutually held values, and exploiting available talents.

At this point, it may be a good time to revisit your story arc.

Paige's story

Paige has probably been telling and retelling her story. Her wish is becoming clearer, containing more detail with each retelling. If you are using our Resilience Story Board Template, here is how Paige may have completed this chapter's entry (Fig. 5.3).

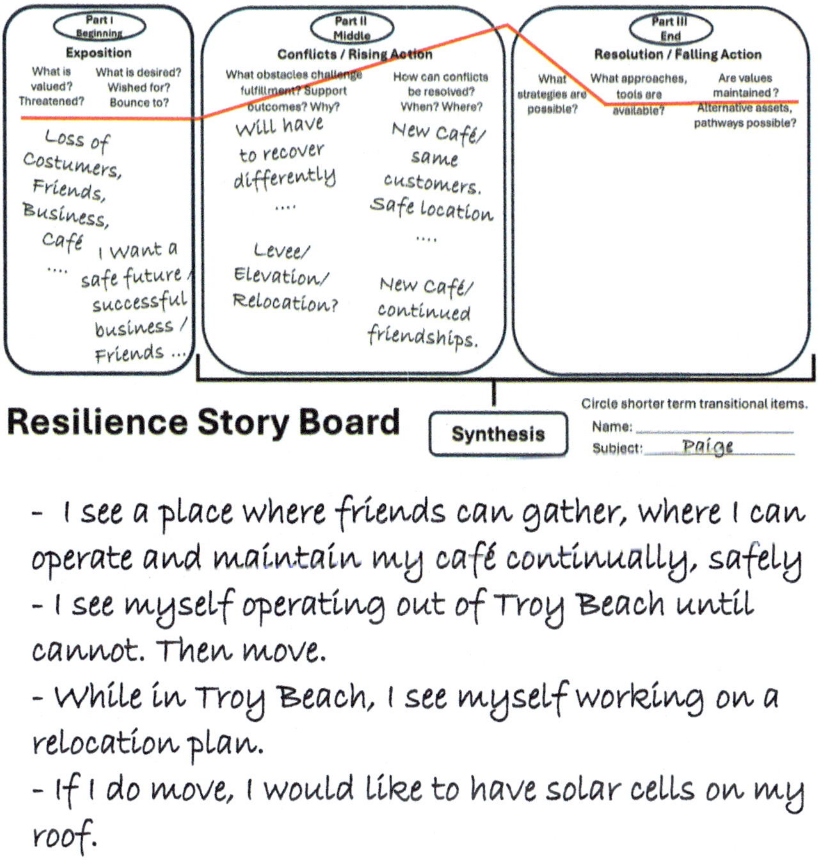

Fig. 3.3 Conflicts are resolved. Paige reimagines her future. Prepared by Bob Freitag

Part III
Resolution/Falling Action

Chapter 6
Celebration

How wishes and outcomes achieved, *revealed*. The Challenge—Will knowing HOW enable repetition elsewhere?

Abstract Watson enters the church where the community will celebrate their success. All is revealed. Those directly involved explain how such a diverse community could achieve common ground. This chapter presents the "Reveal" in our *Resilient Shores* mystery. The who-done-it and what-was-done were presented in Chap. 2 and revisited in Chap. 5. Why it was needed was presented in Chaps. 3 and 4. This chapter presents The Reveal as a "HOW" it was accomplished. You are asked to state how your characters become resilient. For the storytellers among you, was there a reveal in your story, or did awareness gradually build?

Keywords Climax · "How" · Plans · Strategies · Thresholds · FEMA alternative projects · If/When, Then, Then · Community corporations

New Troy Beach Setting, Church Sanctuary, Celebration—25 Years After Hurricane Apollo

B. Freitag, *Resilient Shores: Adapting to Climate Change Through Storytelling*,
https://doi.org/10.1007/978-3-032-09047-8_6

Success is recognized

I entered the church and saw L^2, already seated at the back of the sanctuary. He was the contract planner who helped develop and gain approval for the strategy Troy Beach devised so many years ago. I joined him in the pew, and we exchanged greetings. Shortly after, a younger woman joined us.

The sanctuary followed a simple plan. The congregation sat theater-style, facing a raised stage containing the pulpit to the left and three sets of choir benches. One set faces the congregation, and the others are along the sides. The altar and a piano were prominent. A small movable lectern stood at the center of the stage, looking a little out of place. The use of wooden beams and the many woodcarvings gave the sanctuary a warm and accepting feeling.

L^2 told me that the church had received energy-saving upgrades. Except for the glass covering the original leaded windows, the retrofits were not readily apparent.

I was tired from my journey, but this was overshadowed by my excitement. It helped that sitting on the flat hardwood pews proved sufficiently uncomfortable and I had no problem remaining alert and focused. Looking around, I tried to put names and memories to the faces of old friends and acquaintances as attendees took their seats.

Angela was one of the first to enter. I had heard she was Senior Pastor, so I was not surprised to see her wearing a pastoral black robe. Upon entering, she began greeting others. I never tired of watching how her greetings play out. She would come to an acquaintance, reach out with both hands, say a few words, and slowly draw her willing participant into a welcoming embrace. And, just at the correct moment, she would tilt her head to the side so as not to puncture her intended target with a protruding hatpin. I never saw the process unappreciated.

The front row was reserved for speakers. I was surprised to see Ezra take his seat alongside Angela.

Cassandra arrived, accompanied by Cap, and sat in the pew in front of us, not in the one assigned to speakers. We smiled and nodded at each other. Before she could say anything, however, I was distracted by Sherl entering the sanctuary and unexpectedly, taking a seat in the speaker's row.

Welcome:

After all appeared to be seated, Angela walked to the lectern, avoiding the pulpit. She waited a few seconds, then began speaking.

"Therefore, whoever hears these sayings of mine, and does them, I will liken him to a wise man who built his house on the rock: and the rain descended, the floods came, and the winds blew and beat on that house; and it did not fall, for it was founded on the rock. Matthew 7:24 to 29

"Our church envisions an exceptionally optimistic future for all of us," Angela continued. "What we have accomplished demonstrates the success possible from our collective efforts. We are all to be congratulated for our new beginnings. To a life without the extreme hardships, you and your families have suffered.

"We welcome you all to our humble church."

She paused and said a few words in acknowledgment of all who died as a result of past hurricanes. Angela also asked that we remember Belle, Gilbert, the Admiral, Ms. Oh, Mrs. Eniceo, Mrs. Ala'ilima, Raj, and the many who had passed but to whom we all owed so much.

She then called for a silent prayer in remembrance. She stepped back and, after a minute of silence, returned to the podium.

"We are here only because our new City Hall is still under construction. When completed, we will have a new modern building that will be safe from rising seas. We all thank FEMA for allowing funds initially intended to repair the old hurricane-ravaged building to construct a new, safer place to gather and conduct community business. We thank the city staff and all of you for providing guidance.

"There are many to thank this evening. That is why we are here today. To remember how and why we got there. To celebrate our actions, and each other. I would like to ask Ozell, our mayor, to introduce our speakers."

"Thank you, Reverend. Remember when most of us believed that storms just come and go?" Ozell said while approaching the podium. "There were terrible years and good ones. Years our fundamental values were threatened, and for years all would remain normal. For many of us, our value was enjoying our beach and appreciating the opportunities the beach offered and our coastal climate. For others, it was being secure and having a friendly place to live out our lives. Raise our children.

"Fortunately, we realized our climate was warming, our world was changing. Our communities were heading in the wrong direction. We came to understand we needed a new vision. Instead of pursuing a false course, we devised a new path forward. Not all accepted this new vision. We placed implementation on hold. We established targets that all accepted. Thresholds that, when crossed, would trigger action.

"I would like Sherl, who has become a friend of the city these past twenty-plus years and has very similar concerns in his hometown to come forward. He will reacquaint us on how we inched along to a realization of what was emerging around us. How this process gave us time to revision our community. Time to develop a plan we could all agree on and implement. I would also like Ezra to comment. Ezra lives in Valley and is also a friend of the city. Many know him from his work with Volunteers Active in Disasters."

Action—IF/WHEN Needed:

Sherl went to the lectern, carrying a handful of notes, introduced himself, and began speaking. Initially, I couldn't understand why Cassandra was not among the speakers. Perhaps Sherl and Ezra wouldn't be seen as having an agenda. They were both sort of outsiders to Troy Beach, though familiar with the city and witnesses to the process. Angela and Ozell might have thought they would be more objective and less likely to tout accomplishments. Sherl was clearly an outsider trying to look for ways Troy Beach's accomplishments, false starts, and failures might benefit his hometown. Ezra was known as a critic of the need for what was being accomplished.

"You will remember that it took many years following Hurricane Apollo for you to agree on a new pathway forward." Sherl began. "A revised vision that honored your values while addressing climate threats to those values: having a safe place to live, work, play, and raise a family. A beach to enjoy. It was no minor accomplishment to pursue a process acknowledging that many of you believed these threats were unwarranted, overstated, or for some, even nonexistent.

"After you had a year or so to recover from the destruction caused by Hurricane Apollo, many of you took the opportunity to participate in a series of city-sponsored workshops. I observed this public process and initiated a similar approach when I returned home. But I was honored to be allowed to return often and remain involved as you moved forward. To be a sounding board for many of you.

"After much informal discussion and debate, mainly at Raj's, the city was comfortable with holding formal public meetings and offering alternatives, one of which proposed that future growth and post-disaster redevelopment be directed to the Hilltop Community—IF and WHEN needed. These conditions placed on implementation were vital to your approach. You weren't ready to invest in a future when there wasn't a consensus that the threats were certain. The compromise was to start planning, but not implement any strategy proposed by the city. Resources would be provided only IF and WHEN climate threats were accepted.

"WHEN risks exceed your safety tolerances, THEN you could assess the severity of the threats, and only THEN, direct precious capital toward implementation. To prepare the Hilltop community to receive Troy Beach migrants.

"Does everybody remember that the people who had to move from Troy Beach were called Beachers?" He looked at the audience and saw lots of heads shaking and accepting looks.

"Initially, we planned to use the more legalistic word 'migrant' since they would be migrating from essentially Troy Beach to the new Troy Beach. Someone whose name may or may not be Cassandra didn't like that. 'It might be technically true,' she said, 'but they're only going a mile or so, and it sounds like they're going to a new country. They're almost like a club of people. Let's call them Beachers.' That so many of you became. If you look hard enough, you can still see a 'Welcome, Beacherssticker around in the Hilltop area. Much nicer than 'Welcome, Migrant'". Sherl glanced at Ezra and motioned him to come to the microphone. Sherl remained near the lectern.

"Watson," L^2 whispered as Ezra approached the podium. "This delaying action was painful for much of my staff, but it did buy time."

"Watson, L-Squared," the woman next to us announced. "I am Allison. I was a high school student at the workshops. Do you remember me?"

"Of course, I remember you, but I didn't recognize you! What are you doing now?" I asked, but before she could answer, Ezra began talking from the lecture.

Ezra began speaking. "I fully understand why many of you didn't accept climate change back then. This need to act, required considerable debate. What emerged, as Sherl just mentioned, was an approach designed in recognition that many of us doubted the urgency. Others feared what a changing climate might quickly bring.

"IF we reacted, and action was NOT needed, people would be hurt. IF developing the Hilltop community to receive migrating populations were done too quickly, more people would be hurt. The city had to make every attempt to reduce, to the extent possible, adverse impacts caused by action or inaction".

Allison turned to Watson and L^2 while Ezra had paused, and said softly, "Change was already occurring while planners were working on their contingency plan. It should have been begun earlier."

"The city had proposed a kind of insurance approach," Ezra continued. "I, along with many of you, had no problem with this approach. If there was no or only minor climate impact, THEN there would be no claim. Troy Beach would not have wasted valuable resources, and few would have been hurt.

"All of this planning took time. I still do not accept that humans have caused the climate to change. However, change was occurring. That was made obvious as thresholds we established were crossed. Our friends and neighbors were hurting. This doubt allowed for more time for discussion and debate. All will agree that this added time was a benefit. This added time enabled us to more fully understand what was needed. To plan for an organic transition as our city planners liked to say.

"You all have to put yourselves in the mindset of someone living twenty or so years ago. Many of us distrusted the government and what its representatives were saying about climate change. One of the most important questions with the IF/WHEN, THEN, THEN strategy was how to address the WHEN." WHEN would we go from planning to actual implementation, building a community to receive Beachers?

"We all knew when our house was flooded. Our house got wet. When it flooded too often, and there were signs that this level of flooding would continue and redefine the old normal, it would be time to act. However, we all felt that there was need to develop some specific standards for WHEN we should take action. We just needed to define what would be too often. We needed thresholds we would all accept".

Ezra faced away from the lectern for a moment, took a breath, then returned. "Before I leave, I need to give some credit to Gadsden. You all know him as our High School Math teacher. He is not here. He hasn't bought into the community's reaction to our threat, but early on he introduced us to this helpful conditional statement we all, now, know all too well. The IF/THEN qualifying approach. If something does happen, then we take action. He also continually reminded us that if nothing happens then we do not need to do anything. I wish he were right and nothing happened.

"Just thought he needed to be recognized. It took all of us to get to where we are today." He made a real contribution, and it may be why we are here celebrating today.

Angela stood up and remarking in a stage voice, said, "It needs to be noted that we reframed this IF/THEN offering by Gadsden. We added a WHEN to the statement to reflect that it may be more a matter of time, not IF. We also added a second specific, THEN. Thresholds that would trigger action. WHEN the THEN was triggered, THEN we would act."

"However, as they say, the devil is in the details," Allison whispered as Ezra stepped back from the lectern. "The city Planning Department was asking us, citizens, including the Hilltop community, to conduct one big 'thought experiment' with no need for real-world implementation. At least not right away. The problem was that climate-caused changes were everywhere. For all to see."

"I know," L^2 replied. "We stationed staff at Raj's to discuss questions of risk and the need for immediacy."

"It all depended on individual tolerance levels." I interrupted while turning to L^2. "And, like most things, it depends. We're all different, have different appetites for risk."

"We've had existing processes and standards that had functioned as thresholds for years." Sherl said, retaking the podium. "As many of you know, when a flood caused over fifty percent damage, the city called the structure substantially damaged. THEN building codes governing new construction are triggered, and THEN repairs must follow the same codes that applied to new construction.

"What I am saying here is that this IF / WHEN, THEN, THEN strategy was not new.

"Most of you may know that we decided to use extreme tides, or King Tides, as many of you call them, as the threshold triggering implementation. What you may not know was that this use of King Tides came from discussions with members of the Union of Concerned Scientists, the UCS.

"Building off the UCS, and the resulting discussion with many of you at Raj's, it was determined that residents would accept implementation WHEN being restricted by surface water becomes the normal condition.

"There were also precedents for this. The NFIP defined a condition of flooding as when normally dry land becomes wet. Their regulations applied to certain flooded land and not to land that is not flooded. Our only problem was to develop thresholds for normally wet and normally dry. Thresholds acceptable to Troy Beach residents.

"As many may remember after extensive debate, the vast majority of you agreed that WHEN a structure became engulfed by water two times in one month, on average, the land was not to be considered normally dry. We spent a lot of time debating this. A decision was not easy and you all are to be congratulated."

Sherl paused to assess the audience's reaction, before returning to his seat.

L^2 leaned toward me and whispered. "Notice how Sherl avoids using the term 'flooding here.' That was a problem for our planning team. If the normal condition is being wet, and flooding is defined as land that is normally dry, not wet, you cannot say this now normally wet land is 'flooded.' It is a pond, a lake, or an extension of the ocean. It is best not to refer to this condition as flooding. Flooding implies probability. The probability of when normally day land will become wet. You just can't say land under a lake is flooded. It is confusing.

"This threshold of normally dry land becoming wet twice in one month was actually realized for structures closest to the beach within five years of the workshops. You could see the debris line formed by the ordinary high-water creep under the elevated homes. Land was removed from its floodplain designation to one describing submerged land.

"Crossing this threshold became the THEN when the appropriate elements of the strategy would THEN be implemented."

"We still have flooding." Allison added. "Probabilistic changes to this evolving new normal. Flooding to lands that had less of a chance of becoming wet before the sea levels increased. I remember this being explained by both of you at Raj's. Flooding atop sea level increases."

"This was all evident well before Hurricane Apollo." Allison said while Sherl returned to the front pew."

"You make some great points," L^2 said, leaning over me to respond to Allison directly. "As you made them when a student, but you also must acknowledge that communities usually evolve slowly. The delay caused by those like Ezra, who questioned the science, was exploited by us. The added time helped us design a more human-scale vision.

"This also allowed us to discuss various IFs and WHENs. Alternative environmental changes, projections outside of our control, and our reactions. Different paths demanded different thresholds and adaptation measures. You may hear of this at the potluck they are holding in the basement after this more formal celebration".

"Scenario planning." I said, remembering various scenario planning methodologies I had experimented with, with my students.

"This discussion of timing, triggers, and implementation was important," L^2 said, turning his head toward me. "You need to address it in your account. It is important to note that Troy Beach residents were forced to leave the city, but this was NOT due to some policy decision. Sure, there were triggers and thresholds, but they left because a changing climate was causing seas to rise. It was not the result of some aspirational hidden agenda. Stress this in your book."

Cassandra, hearing our whispering, looked over her shoulder to face us, and smiled. We immediately recognized both her acceptance of our opinions and her tacit disapproval of our interruptions.

Governance:

Ozell returned to the lectern.

"Until recently, before I was elected the mayor of Troy Beach, I headed our Community Redevelopment Corporation," Ozell said. "The CRC. Our goal back then was to ensure there was equity. That resulted in a transition benefiting both Beachers and our existing Hilltop community.

"Historically, we did not have a lot of trust in Troy Beach, and I was told that many Troy Beach residents actually feared those living in Hilltop. To establish a unique special purpose organization for WHEN, we needed something like that CRC. We thought it the best way to create trust between the two communities. We made sure that the Board included representation from Hilltop as well as Troy Beach. We also had representatives from small business owners, residents, developers, tourist interests, and elected officials.

"The implementation of the WHEN in the IF / WHEN, THEN, THEN strategy," Ozell continued while turning to Sherl. "The approach, as you may recall, was reliant on us trusting each other. The CRC was created with little expense following the workshops. And, it lay dormant until the thresholds just described were crossed. We must thank FEMA for a pilot grant to explore the concept and create the organization.

"I have heard that many now feel it may be time for the organization to be incorporated within the city governing structure. But that's a question for another day".

Ozell returned to the front pew.

"Ozell has changed. He's different," I said. "His speech, posture. He sounds like a planner and a politician. The good kind. But not like the tradesman I remember. More serious, more formal, although he always took his tasks to heart."

"His change may be more with you than him," L^2 whispered. "He has been working closely with Cassandra and has been coached by her all these years. He has also become a nationally recognized celebrity for his work in Troy Beach. You know the idea for the CRC resulted from a conversation between Cassandra, Ozell, and Ozell's son. His son was practicing law out of state then, but he provided us with a lot of legal advice. As a volunteer at first.

"His son thought there would be too much pressure from Troy Beach and development interests to prevent or deflect the migration. His thoughts were that Hilltop was going to change anyway, and it would be best to take the reins early. Also, according to discussions with Angela and Ozell, the Hilltop community had not prospered, and they could use the evolving situation to change that."

"It is probably safe to assume that neither Hilltop nor Troy Beach entirely trusted city government," I said.

Overhearing the interchange, Allison leaned in my direction. "The Hilltop community has a long history of being exploited, and the Troy Beach residents have, well, a healthy appreciation for being independent of government and able to buy land when and where they please."

"Most must have gained sufficient trust," I said. "They approved the creation of a unique Redevelopment Corporation. Building trust was a key reason Cassandra held those workshops."

"I think we are certainly better off for having held them," said Allison. "As the process progressed, participants became more trusting. Of each other. I know I have. After all, Ozell went from being a handyman to heading the corporation and then to mayor. Now, the CRC may incorporate within the city. I do not feel this would have happened without these workshops and the intense discussions that followed. Over the dinner tables and at Raj's."

"I welcome you all to celebrate our shared progress," Angela said, returning to the podium. "To celebrate our future, IF / WHEN you want to speak." Angela said, grinning at her word selection, "Please raise your hand. We have a few student interns from Tougaloo College and State with roaming microphones. They have been assigned to our CRC to help with this ceremony.

Celebratory Statements:

Allison stood. An intern passed her a mic.

"I want all of you to recognize how we are addressing all of our energy needs. I was a high school student when the city held its workshops. After graduating from State University and interning with the CRC, I then got a job with the city. I am now the City's Resiliency Officer.

"Troy Beach is becoming energy positive. We will create more energy than we use, and with this excess energy, we can remove greenhouse gases from the atmosphere. To create energy, we are building solar farms outside of town, we have installed wind turbines offshore, and are working with State to distribute floating tidal/storm mats along the coast to capture wave energy. The city will be carbon-negative. We are building with carbon-absorbing materials and carbon-friendly practices. Our city layout protects us from hurricane-force winds. Its design also routes cooling breezes through the city. We are taking advantage of the cooling properties of street and vertical gardens, making our city green. You've also, I am sure, noticed the cooling water mist along critical corridors. We are using smart technologies, and energy is not wasted.

"Energy storage was a problem. We had planned to build a water storage basin outside of town and pump seawater to the reservoir when energy was abundant. Then, the water would flow down through water turbines when energy was needed. Thanks to new battery technology, we could scrap the project.

"And lastly, thanks to the courts and a group of young plaintiffs, fossil energy companies have had to build carbon-sequestering plants. You have just heard from Ezra. He helps run our local plant."

Allison walked away from her seat, and before returning and handing back the mic, she added. "Fortunately, the facility could be located near an abandoned pipeline and could collect transported emissions. It also employs direct air capture. Initially, carbon was pumped underground. The court case required that they sequester carbon, but the court did not say they could not make a return on their investment. Currently, they can store carbon as a construction material. And sell the product at a profit.

"It figures," Allison said quietly while relinquishing the microphone. She looks at me and says, "They always win. But it's not a bad thing. It's a win-win. The process is called the 'Technological Recovery of Environmental Emissions' or T.R.E.E.S." No irony there."

Atlas stood and was given a mic.

"As Allison said, we are using this new construction material. For those who do not know me, I am Atlas, and I have been a contractor in the area for decades, since before Hurricane Apollo. Back then, Cap owned the Bargain Beach Wear store and almost got me arrested for constructing without a permit." Atlas said while looking at Cap and Cassandra and, with a touch of drama, putting on his old bucket cap given to him many years earlier.

I looked at L^2 and Allison, and like clockwork, all three of us reached into the bags we brought and put on identical hats.

A few others in the audience did as well. The rest either laughed or looked perplexed. Depending on how long they'd been around Troy Beach.

"But, actually, thanks to Cassandra and Cap," Atlas continues. "I discovered a new vocation and market."

"Our new Troy Beach needed housing, and Gilbert, who worked for the city then, and I designed a modular construction system. Gilbert returned to State and developed the approach as his required PhD dissertation. He passed on," Atlas pauses to collect his thoughts, "a few years ago, and I want to thank Angela for the remembrance.

"One of the reasons housing was so expensive was that construction was on site, and materials were of all sizes and shapes. Housing components were shipped here from all over. Nothing wrong with our local architects, but each seemed to try to outdo each other in creativity and uniqueness. Each project employed different materials and construction methods. We hired those same architects and asked them to create designs that could be built as modular elements. We still wanted interesting, creative human-scale structures, but cheaper. In many cases, it can be dismantled, relocated, and reused. In other words, mobile.

"They accepted the challenge and designed buildings that could be manufactured off-site as modules. And as I said, and should be emphasized, one key criterion was that these modules could be disassembled and reassembled in different locations. And, to the extent possible, using the recycled carbon materials made available through T.R.E.E.S., our carbon capture plants that Allison mentioned.

"The new multi-unit development down by the shorefront offers an example of our approach. The site is at the edge of the FEMA-designated floodplain, and rising seas and higher tides are not yet inundating the project. The city estimated the site has about twenty years left before crossing our 'normally wet' threshold. We can easily build structures with a useful 20-year life that profit their owners and bring pleasure to their occupants.

"I also want to mention the pickleball and accompanying tennis courts intended to be constructed behind the development. Nothing to do with flooding, but as most of you are aware it gets really hot in the summer. The active areas will be sheltered with balloon-looking positive air pressurized covers. The covers are made of a material that will actually generate electricity to be used to keep the courts cool. Dry in winter. During storms, the covers are designed to be deflated. We are really proud of these and have received interest from many communities. They are not excessively expensive and will enable year-round use."

Gilbert paused, looked over the crowd and continued. "Initially, we couldn't locate workers. Lynn, the academic extension director with our State University, worked with Thatcher and me, and together, we wrote and are now delivering a curriculum on modular construction. The YMCA, under contract to State, helps deliver the program locally. I employ many of its graduates.

A well-dressed, older woman stood up and asked to speak. Atlas called on her, with a smile for his old boss.

"My name is Thatcher," she said, taking a microphone. "I was a developer in Troy Beach when this little coastal city was booming. I am retired now, and my son has taken over the business.

"This approach Atlas introduced did not prove to be as arduous as you might think," Thatcher said. "There was profit to be made. The units built off-site were designed to be assembled on-site, and when thresholds were crossed, they could be disassembled, moved, and reassembled at a safer location. Space is being provided in Hilltop to accommodate the transfer if the owners are interested. Again, think of large Lego blocks. Plug-and-play units that can be physically relocated as needed.

"I also need to thank Troy Beach Bank for making it easy for purchasers to get mortgages for these units. I am sure Atlas wants to thank the high school, community college, and energetic YMCA counselor for providing skilled workers. When we started, there were too few workers in the building trades, and no one was trained in modular construction, as Atlas just related.

"As for me, I worked here during the boom years. Before Hurricane Apollo. Back then, anything we built would sell.

"Life was good, and my business was profitable as long as everyone ignored climate warnings, that is. Following the workshops held a few years after Hurricane Apollo, I realized my business model had to change. The workshops increased awareness, but the real change came when thresholds were crossed. When one hurricane caused a storm surge that undercut the foundation of one of my condo developments, the structure had to be abandoned for safety reasons and was demolished. Actually, the city worked to get that hurricane declared eligible for FEMA disaster recovery assistance, and the development was determined debris under the Public Assistance Program. Removal was paid, in part, by FEMA.

"Ozell, when he heading the CRC, convinced me to work with the organization. The playing field had changed. Hilltop property owners often wanted to partner and not sell their interests outright. There were restrictions, and I had to buy density bonuses to maximize my return. This included purchasing development rights, or entitlements as they are often called, from at-risk Troy Beach property owners. Properties are located in what the city designated as sending areas. They were applied to what the city had designated as receiving areas within Hilltop.

"All of this took time and money out of my pocket, but the CRC helped me negotiate the process. Serviced land, which had water, sewage, and power available, was in short supply. Not like in the old days.

"We still work with property owners, hire architects to design the building, and contract with Atlas to supply the structural units. Most structures are built from modular elements now, even printed in some cases, and the site is prepared to accommodate these modules. As Atlas mentioned, they are also designed to be able to be disassembled economically."

Angela rises from her seat at the front pew and hands the mic to Ezra, who indicates he wants to comment.

"I need to add here that the CRC is providing affordable housing. Many of our past residents are no longer low-income because they have been able to partner and profit from their association with these developers. But there are still people in need. Beachers and Hilltoppers alike.

"The approach that Atlas and Thatcher developed has cut construction costs considerably. Some of the funds received from developers buying added density bonuses are used to purchase sites for lower-income housing. We created a Housing Trust as part of the CRC to coordinate the effort".

"Richard?" Angela announces while seeing the FEMA officer requesting that he be given the microphone.

"I'm the FEMA Regional Administrator and I need to comment on what Thatcher mentioned and you, Angela, touched on in your opening statement.

"After the hurricane and Disaster Declaration that followed Apollo, and at the direction of Troy Beach, we diverted considerable FEMA disaster recovery funds to what we call Alternative Projects to the Hilltop community. These were eligible awards that would have gone to help the old Troy Beach recover. This reassignment was no small feat, and your celebration today should emphasize this. Recovery funding estimates, based on losses in the functionality of existing eligible assets, were diverted as Alternative Projects to help restore lost functionality, only at a different location. Hilltop."

"As all of you should know by now, being one of our best customers," Richard paused, smiles, and looked to see if his attempt at mild sarcasm was acknowledged. "FEMA helps communities provide temporary housing, repair damaged housing and public infrastructure, and reimburse residents for unmet needs. If a loss was the result of a FEMA Declared Disaster and meets our requirements, we can assist with recovery. We can also add additional funds to projects when the extra funding may reduce the likelihood of future losses or support new codes and standards.

"However, what happens if or when the place where the damage has occurred can no longer support safe development? Recovery becomes impossible. This was the case for Troy Beach.

"It is always easier to repair than relocate an asset. Houses, churches, and roads… must be seen as elements in a system. All are interdependent.

"Your experience here, in Troy Beach and Hilltop, is providing an example that a change in direction is possible. Not easy, but possible when the planning foreshadows an event. That's why your IF / WHEN, THEN, THEN approach was so exemplary…" Richard pauses, "and practical.

"I am also proud that FEMA helped fund your CRC.

"But it is you who should be celebrated. FEMA is just a checkbook. We provided technical assistance when asked, but we're just a checkbook with guidance and regulations. We can also be cheerleaders, and I guess that's my role here today. You and your planning upped the value of what FEMA could support. What we helped you create with our collective public tax dollars."

"Nicely said," L^2 said quietly, looking in my direction. "Sometimes the press presents FEMA as a knight in shining armor, riding into town like some kind of Mighty Mouse singing, 'Here I come to save the day.' While all they may have is a checkbook, they can advance the client community's aspiration."

Allison glanced at us and whispered, "Who is Mighty Mouse?"

"Don't think FEMA owns a single shovel, broom, or hammer," I murmured to L^2.

I leaned over the pew and told Cassandra that her approach not only provided time to plan, it also allowed time to build trust and acceptance.

My conversation with Cassandra was cut short when Gadsden stood and was given a mic.

"I need to say that all is not all wine and roses," Gadsden announced.

Angela's glance at Gadsden displayed apprehension. Gadsden continued unfazed. "I inherited a house down by the beach. Every once in a while, my home was damaged by a passing storm. I would gather debris I salvaged along the shore and repair my house. Some five or so years after Apollo, my home was damaged again, and the City offered to buy it with FEMA funds.

"I did not want any government agency taking my home, and I refused. Seas kept slinking inland, but I kept up with the problem by adding fill under my elevated home. Then another storm came and damaged my home, then another. Some were declared Presidential Disasters. Some not. However, more importantly, the shoreline was migrating farther and farther inland. Farther under my house with each storm. After a few years, my house was one of only a couple of houses remaining on the beach. I was becoming the sole survivor. I had crossed the regulatory threshold talked about much earlier. Troy Beach city government did not attempt to repair the beach. Repair the dunes. Build a sea wall. And the FEMA Buy-Out program had ended.

"My property became an island. Often a submerged island. I was not allowed to rebuild and occupy it after a storm. My house was demolished.

"I didn't get any flood insurance claim because I could not afford a policy. FEMA gave me some 'unmet needs' Individual Assistance money, and Thatcher bought some development rights entitlements. But the money I would receive was nowhere near what I would need to buy a new place. I had to live with Ezra for a while, but finally, I was able to buy a home in Valley. It is small, and it floods, too.

"I could have afforded one of the homes in Hilltop, but they were all new and multifamily. I just did not feel comfortable in a building where I could not scream if I wanted to maintain my home myself. Paint it the color I wanted. Hammer a nail in the wall if I wanted to. I did not want to live where men like me are minimized, made to feel guilty, and the cause of all that's wrong with the world. In Hilltop, I could not live the way I wanted with the people I wanted to live with.

"That's all I have to say."

Angela saw Manisha raise her hand and acknowledged her.

Given the microphone, Manisha stood, ignoring Gadsden's statements addressing the gathering. "I am Manisha, Raj's daughter, and I manage the Chang'en Cajun. I just want to say a few words about my dad. Most of you know he helped with the tram line many of you rode to get here, but what I am most proud of is his creation of what Mrs. Eniceo branded as the Dirty Kitchen. Not a name I would have chosen, but she laughingly called it that one day and the name stuck. Cassandra probably came up with the idea, and my dad ran with it.

"The Kitchen became so much more than a place to cook and, in many cases, to consume a meal. Cassandra said she wanted to build a mutual aid network and place of synthesis. I have no idea what that meant. My dad just saw a need and wanted to help.

"Sure, many needed a place to cook when the power went out. But if you needed a hammer or some help to wield that hammer, you would find it at the Dirty Kitchen. Resources were shared, and problems were solved. The ideas presented here by Ezra and Sherl, like the meals, were brought to the Kitchen, displayed, cut up, blended, simmered, consumed, and digested. A visiting medical professional came to the Chang'en Cajun weekly. This idea too arose from discussions at the Dirty Kitchen.

"All was not serious. Hilltop men started a poker group at the Kitchen. They called themselves the 'Cheaters.' Ten dollars per evening limit, and one bottle of beer per evening. It was not only for men. Often, the women wouldn't even play for money, but 'servitudes', as they called it. Belle bet Ozell a mango pie to repair her outside door that hadn't been able to close properly for generations. Belle lost. Ozell got the pie. Many of the Cheaters enjoyed the pie and Belle's door, mysteriously, got repaired.

"The Dirty Kitchen, as it was, is closed. It is not needed now. We have the Belle Community Center as our gathering place, and we have a more reliable social infrastructure, as Cassandra says. The Center has a new kitchen and eating area. Also, the CRC posts the status of each project at the Belle Community Center and provides for a weekly discussion on community topics. Twenty or so years ago these kinds of discussions took place at our Dirty Kitchen. The events are now streamed live.

"But the Dirty Kitchen is remembered. Many of you have dined at what was the Dirty Kitchen and may not have realized it. We covered the site with energy-producing glass, where you can dine amongst the vertical gardens. The old motel, with its simple, gentle curves, balconies, and courtyards, helps frame this indoor park. Our menu is built off our produce. Tomatoes, herbs, squash, lettuce… they are all grown on-site. We also supply other Troy Beach restaurants. If you look closely as you walk about the gardens, you will discover an old Dirty Kitchen picnic table. I want to thank Dad, who also helped with that."

I heard some giggling, more like chortling. Behind me. I was pleased that the commotion was softening the mood, but it had become a little disrupting. I turned and saw a gaggle of pre-teen girls. Behind them was a tight cluster of their male classmates. I had taught middle school students and was very familiar with this syndrome. Some event, amusing to no one else, would trigger the funny bone of a single girl, and once the caged emotion was released, like some genetically inspired visceral explosion. There was no stopping the chain reaction.

I looked at the boys, nestled together. All were eyeing her with a smirk that I, as a past teacher, had become all too familiar with. With all of the concentration I could muster, and tried to project… Please, I beg you, don't let gas. Don't fart. Please!

I felt sorry for the tween whose hand was pressed tightly over her mouth and was trying to quell the emotion. Her friends offered no help and were prodding her to raise her hand. The boys, sinister beings, were not her allies.

Angela shot the children look that transmitted disapproval, concern, and sympathy all at the same time. She pointed for a microphone to be taken to the guilty party.

Recalling when I taught middle school, I instinctively looked over the assemblage of embarrassed students and, without realizing it, gave a stern stare at the assemblage, daring a member to continue the turmoil.

"Keshi'a, would you like to contribute?" Angela asked.

The girl bounced to attention. Her friends cowered into their seats.

"Reverend Taylor, I would like to thank you for changing the school year, so our vacation now starts in November and continues through January. I remember how hot it was in the summer, and a lot of times, we couldn't play outside. It's better being in school when it's so hot. The new school is cool inside, and all the trees in the play yard help too."

I had never heard Angela referred to by her last name before. My posture became a little more proper.

"That was a national discussion, but thank you," Angela answered. "Your comment is appreciated."

A person I had never seen before stood and took the mic from an intern.

"Hello, I am the son of the Eden family, who owned a condo and participated in the workshop you mentioned.

"My parents have passed on, and we inherited their property. That is until the seas and storm took it. I think it was the one Thatcher referred to. We got a flood insurance claim payment and a little money for selling our entitlement and our development rights.

"Our family pooled what we had, and we bought a leasehold interest in one of the modular apartments in the development Atlas mentioned. The ones down by the beach.

"We have a lease for twenty years, or until…, WHEN, the structure may be taken by a storm or encroaching tides. Whichever comes first. We are now of similar age to our parents, when they lived here, and we bought the property for our grandkids to enjoy as our grandparents did so many years ago. There is ample warning of approaching storms, and evacuation routes are well-marked. So, we feel physically safe.

"We also have purchased one of the new types of insurance policies that provides some security should our home be taken before the lease is up. Kind of like a life insurance policy. We need some protection if our life in the home is shorter than expected. The insurance company is betting on its use lasting longer. We don't need a storm event to collect. That's why we also have a flood insurance policy. Payment is triggered for the life insurance type policy when we can no longer occupy the structure. When we cross the threshold. Should high seas cause a general condition of flooding and we have damage, we can make an insurance claim on our flood policies."

"The NFIP," I said to L^2, "is considering paying claims when occupancy of insured property is deemed unusable, unsafe. A state of un-usability would result when a property is inundated by King tides twice in a single month. Regardless of there being damage at the time of the claim. The claim payment could be adjusted to reflect an estimated period of usability. Currently, FEMA prepares risk maps that illustrate estimates of when the land will no longer be considered as normally dry. When the threshold is projected to be crossed. The regulatory aspect of the new Coastal-NFIP program would require improved property to be used commensurate with the threat."

"Our units resemble shipping containers and are transportable. We are considering purchasing a vertical plot in Troy Beach," continues the Eden family representative. "We want to see how the city-to-beach trail system works first, but we're thinking of it. We only bought the home to enjoy with our grandkids, so we're not thinking of it as a long-term investment.

"On a side note, the beach is wonderful. It's larger than I remember as a child, and the emerging marshland and mangroves frame the swimming areas nicely. That is, in the winter months. The summer heat is brutal, and we try to avoid coming here in the hot season."

"I'm Rover," said the next man taking a microphone. "I owned a mobile home back in the day. Still do, actually, but mobile home life has changed. I am older now and don't like moving from place to place. The city has a better approach to mobile homes now. They've provided mobile home parks down by the shore, but you must be prepared to move when notified. All supporting facilities can also be transported. The shower building, bathrooms, and solar array can be carted off the site. Some days there is a food truck at the park, and well, it's a truck, so it too can move too. There is a mobile home storage/refuge site upriver beyond the Valley Subdivision that is set aside for shelter during storms.

"The city makes a nice income from us at minimal risk. The temporary nature of the site is okay for most. The summers get hot, though, but there is usually a nice breeze in the other seasons. The mangroves help."

"This is Cap," a man said while requesting a mic. "I must say a few words about our old and new historic district. I have said that idea of a NEW Historic District seems silly, but that is what we call it. I owned the old Bargain Beach Wear Store. I'm sure Paige, our café owner, would agree with my sentiments when I thank you for creating an orderly, appropriately speedy transition. The OLD Troy Beach Historic District seems equally ridiculous. OLD Historic District? Anyway, it was functional for a long time. Well, as long as there were patrons and our establishments survived disasters. The city built a little temporary berm around the district. This gave us about ten years before we crossed the threshold mentioned earlier. Time to plan our move to Hilltop. I also want to thank our local Chamber. They were a help, as were the Lions Clubs. The Rotary and Kiwanis clubs were active in neighboring communities, and they helped too. Their effort got us all on the same wavelength. Our actual move started only a couple of years ago, and as of today, we all have relocated to the NEW Historic District, and are functioning." He chuckled. "Meaning making money."

"You are spot on," interjected Paige, who had arrived late and was standing in the back, speaking loudly without a mic. "Timing was everything. We also have to thank the bank for giving us a little grace on our loans. Just thought I'd add."

All of this was because of Cassandra, yet she was not acknowledged. I leaned over the back of the pew and got her attention.

"Cassandra, we're here because of you. I remember, as if it were yesterday, your torment in not being believed. The frustration in that few acknowledged climate change, and even fewer, outside of Allison, wanted to take action. How you created a public process where those impacted could take control."

"Thanks, Watson, you've always been kind. You can make a lot of difference if you don't demand credit. I asked Angela NOT to give me a role this evening. This is their celebration." She beamed with pride.

The more formal part of the ceremony was concluded. Most retreated to the basement for refreshments. Angela left the front of the sanctuary and did what she did best, give acknowledgment hugs to all she met. People left her embrace, wearing her infectious smile.

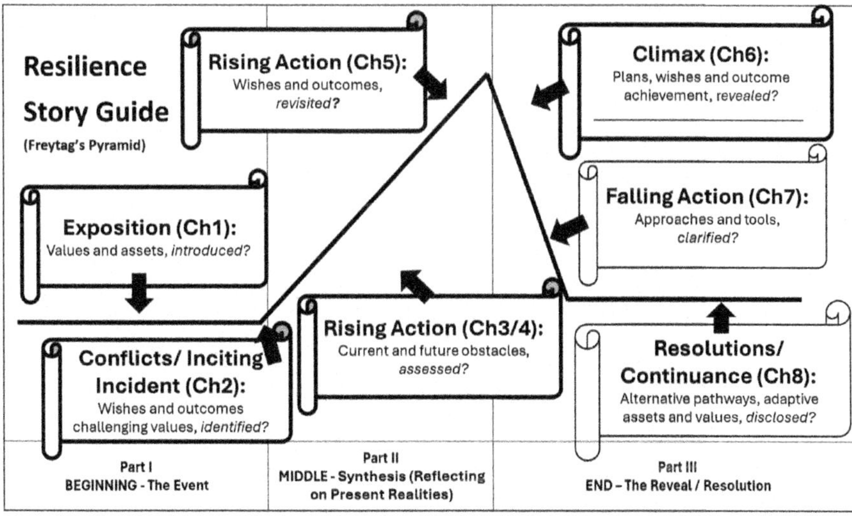

Fig. 6.1 Freytag's pyramid—success is recognized. Prepared by Bob Freitag

Building Your Own Story: Cassandra's Suggestions

Climax/Plans: How wishes and outcomes achieved, revealed. The Challenge—Will knowing HOW enable repetition elsewhere?

HOW we reached agreements is finally revealed (Fig. 6.1). This is the purpose of Watson writing his book.

How did you react to Watson's disclosures? You now know HOW the conflicts were resolved. Thinking of WOOP, you now know the Plan, and HOW our communities' WISHES and OUTCOMES were achieved. You know OBSTICALES confronted, and the resulting PLAN.

If you're writing a mystery, think of this chapter as Watson's reveal. His reveal of HOW.

The values introduced in the first chapters have not been forgotten. The threats to the Beach, life safety, livelihood, and community networks have been addressed. Common ground was achieved, and trust was built along the way. With the rapid onslaught of climate change, implementation was needed quickly. Planning takes time, so we thought framing noncapital investment as a contingency plan would accomplish our goals.

Plot Development/Conflict Resolution/Resilience Concepts

Whatever plot you create, your task is to create a plausible resilience story for some individual or a community of individuals. A resilience story that can provide direction, hope, and passion. And is achievable.

Climax:

This is the most intense part of your story. It is the turning point. It is the reveal if your story is a mystery. The reader/listeners discovers how your characters confronted the conflict.

Questions:

- Do you have a REVEAL? Or is awareness gradual?
- Was there a turning point, a threshold crossed?
- How did your characters resolve conflicts?
- What did you bounce to?
- What emotions or outcomes hinge on this critical moment?

You may have assured the continuance of your values by reducing vulnerabilities through protecting assets, accommodating the hazard, or relocation. Your plot may have involved bouncing back to an agreed and achievable path, or you may have had to re-organize to a new trajectory. Whatever story arc you are creating, you need a story that reduces your assigned threat, provides hope, and generates passion. Here are some thoughts.

Who? You may need some new characters to underscore nuance. Watson introduces some new characters. Manisha, Eden, Richard, Keshi'a. Each helps build a greater understanding of WHY and HOW Troy Beach communities have structured their overall strategy.

Similarly, you may need to adjust or add voices and faces to your arguments when presenting your resolution.

As I have said continually, make sure that your story is not just about an abstract community. Make your characters' struggles personal. And, their resolutions are reassuring, hopeful, and even joyful, if possible.

What? Our three communities, well at least Troy Beach and Hilltop, have crossed tipping points. It would be unsustainable to return to an old normal. Figure 6.2 offers a standard illustration of a community entering a new STABLE regime.[1] The first depressing represents the old Troy Beach where the tidal ebb and flow are within reasonably stable parameters. Climate change has rendered this condition unstable. This boundary is being crossed. New stability can be reached with Troy Beach's relocation to Hilltop.

Angela, Sherl, Ozel, and Ezra have created a celebratory moment. Did it work for you? Are the resolutions being revealed reflective of the hard work and pride undertaken by Troy Beach, Hilltop, and, to a limited extent, the Valley Subdivision?

Has your approach fortified your target's current regime or reorganized and created a new one?

[1] Fernando Jaramillo, "Hydrological tipping points: Can we tip the bucket?" EGU BLOG, 2020. (Figure 6.2 was based on that prepared by Dr. Jaramillo.)

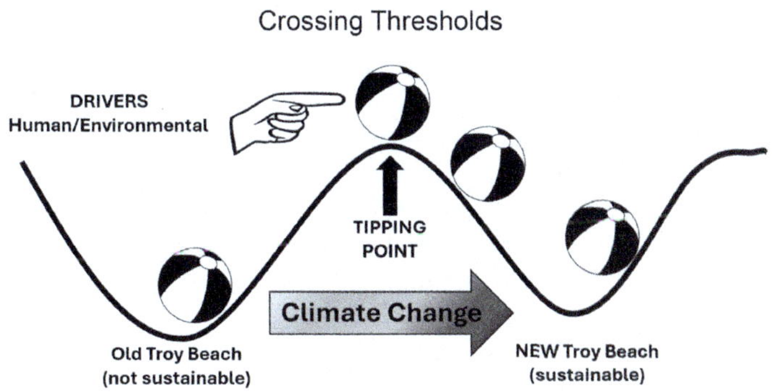

Fig. 6.2 The two depressions represent independently stable systems. Movement to a second system occurs once a threshold is crossed. Prepared by Bob Freitag

When? WHEN and IF contingency planning arguments are given prominence by Watson. Developing a Contingency Plan may not address your needs.

We, urban planners, would have welcomed the opportunity to begin implementation as solutions are determined, evaluated, and agreed upon. We couldn't. We had to build trust first.

If there is trust in your resolutions, you have just jumped the highest hurdle.

Where? The importance of Hilltop increased as our story evolved. If Hilltop had shut the door to Troy Beach migrants or if Valley had marketed its community to outsiders, our resilience story would be very different. Hilltop may be thought of as a metaphor for the many communities, the members of which could not afford shorefront properties and had to live away from the workplaces. Places becoming more valuable as shorefront neighborhoods are threatened.

For fun, consider exploring a variety of resolutions, emphasizing alternative values of your characters, or different expressions of the hazard. We attempt this in Chapter 8.

Why? Why did Paige and Cap take a chance with Hilltop, why didn't Gadsden? Why was Belle ingratiated with the community center named after her? Why does Ezra now work for a "Green" organization? Not all questions have to be answered.

Your story does not need to completely tie up all loose ends. And, it may be more engaging if you give your listeners some raw meat, unanswered conflicts.

How? At last, you now know HOW these communities are organized around a common wish, a goal. A process that relied on staged relocation to safer ground. One that made the time needed to concurrently implement hazards mitigating approaches while pursuing unsustainable direction over the short term.

The HOW has been revealed, or at least some of the more critical HOWs.

Gadsden and Ezra were not the only residents of Troy Beach who questioned climate science. How could the community proceed? These contrarians accepted the concept of thresholds, which, all agreed, would trigger action when crossed. IF and WHEN. This allowed for planning to proceed.

Although the threshold was when normally dry land became normally wet, many had their own personal thresholds. Paige's was when her client friends no longer frequented her café. Cap's was when he had customers. Gadsden's tipping point was when he could no longer repair his home.

After the workshops, residents gained trust in their friends and neighbors, or at least all felt they shared common values. And they were in the same proverbial boat. They questioned their respective governments and created an organization led by those they trusted. And the Community Redevelopment Corporation was born.

Summers were becoming very hot, and Keshi'a provided the HOW that she, even as a youngster, helped change by pointing out the need for rearranging the school vacation schedule.

Other HOWs included, the gathering place offered by Raj and his "Dirty Kitchen," FEMA's use of Public Assistance funds, and staggering, phasing, and adaptation approaches corresponding to a migrating shore.

Paige's Story

Paige was concerned about thresholds and having the commercial community relocate as a group. She has come to see relocation as a process, exploiting each resilience definition. See Fig. 6.3 for how she might make her entry for this chapter on her Resilience Story Board.

It may be helpful to integrate your most interrelated HOWs, now. You can expand upon them after reading the next chapter when many of these approaches and tools will be further explained. If you are not interested in this level of detail, feel free to skip this chapter.

I recently found an approach that may be helpful if you want to experiment with our IF/WHEN, THEN, THEN method, and use of thresholds to change direction. Search under the heading "Dynamic Adaptive Policy Pathways. (DAPP)".[2]

As I have said after most chapters, your story, your process, your plot will differ. It is your story, after all. I hope you had fun with it and that, as you and your friends and colleagues retell it, it gains acceptance and traction in your community as it becomes a reality.

[2] Haasnoot, Marjolijn. Dynamic Adaptive Policy Pathways. All ECSTATIC Materials, Paper 92, 2002 https://digitalcommons.usu.edu/ecstatic_all/92.

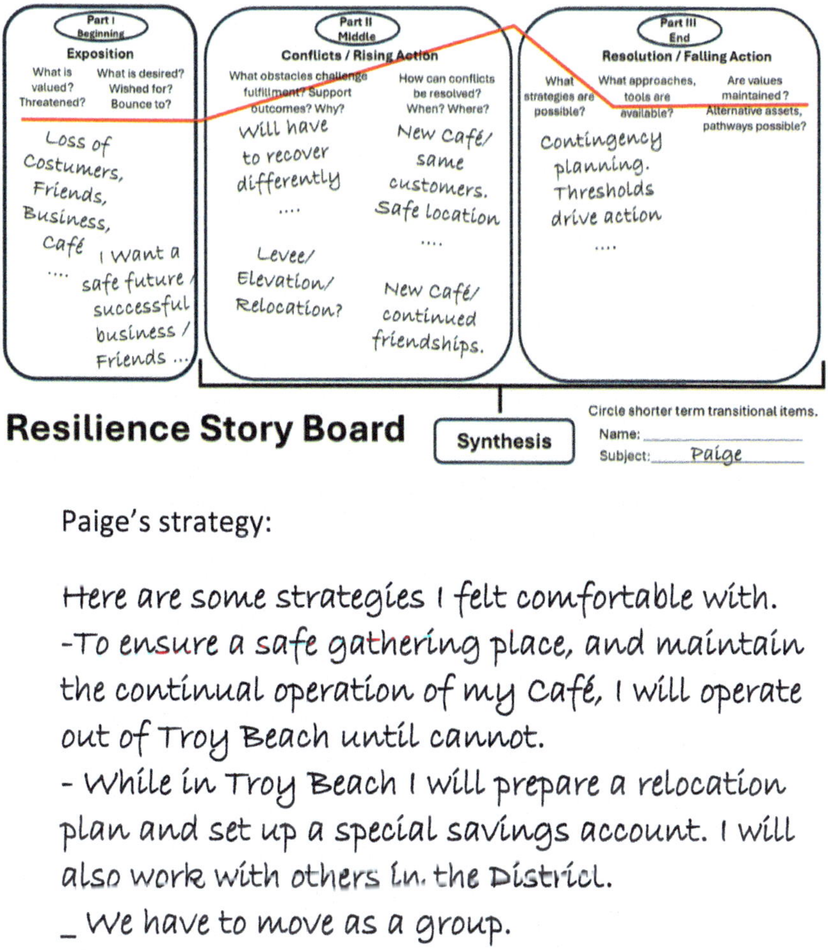

Fig. 6.3 Paige's strategy is revealed. Prepared by Bob Freitag

Chapter 7
A Deeper Dive—Approaches and Tools

Approaches and tools, *clarified*.
Challenge—Limited resources

Abstract Watson stepped outside the sanctuary following the more formal part of the celebration to confer with those having a direct hand in the success. If Watson wanted to chronicle the Troy Beach Story so that others may duplicate its success, he would have provided sufficient detail to allow readers to research the specific planning, partnering, and financing solutions used. Partitioner design professionals and nonprofessionals alike would have to understand the approaches used. The HOW introduced in the previous chapter is explained in detail, from the residents' perspective. The reader is given sufficient information to initiate similar activities elsewhere. The reader/storyteller is asked if their story includes sufficient detail that others may fully comprehend the resolution and could duplicate the results.

Keywords Approaches · Tools · Flood insurance (NFIP) · Repetitive loss · Building permits · Entitlements · Purchase of development rights (PDR) · Transfer development rights (TDR) · Sending and receiving areas · Fee simple purchase · Life estate purchase · Lease purchase and transfer · Regulatory thresholds · Property rights · Mitigation plans · Comprehensive plans · Disasters by design · Capital improvement plans (CIP) · FEMA alternative projects · Tax increment financing (TIF)

B. Freitag, *Resilient Shores: Adapting to Climate Change Through Storytelling*,
https://doi.org/10.1007/978-3-032-09047-8_7

New Troy Beach Setting, Outside Church—25 Years After Hurricane Apollo

This image has been altered with permission of the artist. Artist: Ron Kasprisin.

Approaches and tools are discussed

Before moving on to the community celebration in the church basement, I stepped outside of the sanctuary to confer with L^2 and Allison. If I wanted my chronicle to include sufficient detail to allow readers to research the planning, zoning, and financing solutions used in Troy Beach, I had to first understand how they were applied. I want to be able to succinctly explain Troy Beach's pinnacles and pitfalls. Make it engaging for design professionals and nonprofessionals alike without having most readers fall asleep.

I knew I didn't need to provide a detailed reference guide or a tool kit—a host of academic papers and agency reports do that—but I could present what was attempted in enough detail to enable readers to find resources.

L^2 and Allison could help with this.

When I mentioned my idea to Allison, she advised me to first warn my readers when I dove into the more technical material. I said I would. This is it: Reader, beware.

The Community Redevelopment Corporation:

"L-Squared, was the Community Redevelopment Corporation really needed?" I asked. "Seems like an extra layer of management and regulation."

"If not a Community Development Corporation (CDC), or in this case, the CRC, who would take the lead?" L^2 replied. "Troy Beach and Hilltop had long histories of both support and exploitation. Those old biases were alluded to at the workshops.

How could the 'privileged,' 'effete' Troy Beach residents work with the 'lazy,' 'uninformed' Hilltop community? Valley buried their public swimming pool in the '60s. Didn't want to pollute it with Negros, Black bodies. It is still a sore point among Hilltop residents. It was all about trust. And there wasn't much of that. The workshops helped a lot, but they were only a beginning.

"The idea of creating a CDC, or rather CRC, came from Ozell's lawyer son. It was coldly received by both--initially--but during the course of our planning, Hilltop residents began to feel pressure from development interests. There became a sense of urgency.

"The CRC was given broad powers. It could enter into contracts and partnerships with public and private entities for relocating severely at-risk Troy Beach properties to Hilltop. They could administer the purchase and transfer of development entitlements, apply for grants, provide low-income housing, be a vehicle for economic development, assist in creating commercial and industrial uses, and provide and implement other municipal services and functions in support of relocation. As for the restoration of Troy Beach natural lands, the CRC could coordinate with private and public natural resource interests to acquire and restore lands, and coordinate with natural resource interests on the long-term management of acquired newly formed beach and submerged lands".

"What I do not understand is why the majority of the CRC's board were from Hilltop, or at least had a strong connection with the community," Allison asked L^2.

"Probably just reflected the best candidates. As far as I know, this was not a strategy. It may help to explain the transition to Hilltop as a card game. Hilltop had a strong hand; they owned the land. Yet, those from Troy Beach knew the game; they had developed the rules. I'll ask you, Allison, what would convince you to play the game if you were from Hilltop?"

"I guess if I felt the gains outweighed losses," Allison offered. "If I trusted the rules and those playing the game."

"Exactly," said L^2, " Hilltop residents had to trust the CRC and believe the CRC process would be just and fair. Hilltop felt it had more to lose and fewer options than Troy Beach. Not too sure this was true. But their entire history was one of defense. I am confident that some Hilltop residents felt that it was about time their community got their share of the pie. Others felt that individual wealth was finally possible without the weight of the community's past inequities overwhelming them. I think most felt the community could not fend off climate-related migration and saw an acceptance of the CRC as an opportunity to create a diverse, accepting playing field. Change without losing their identity. If all were allowed to play fairly. Also, as I witnessed, the CRC was mostly a consensus organization. Strategies were discussed until there was agreement. Including agreement in not being able to reach an agreement."

"This is something I can ask Angela when we go to the celebration," I said.

Resident IF/WHEN Stories:

"You have to remember that many did not accept climate science back then, Allison said. I think it may be helpful if your book offered at least a little detail concerning the approaches and tools used by Troy Beach. Even if all readers don't need the specific details, it may be helpful to offer some context. You could tell our residents' stories. Gadsden's trials and tribulations would be a good place to start."

Entitlements—Gadsden/Thatcher:

"Tell me what happened to Gadsden," I said, looking at both L^2 and Allison.

"Gadsden," Allison said, "was one of our longer-term residents. I was in high school when I first met him at one of Cassandra's workshops. He seemed often angry at something. I recognized him and his house when my friends and I went to the beach. He never greeted us, although I feel he must have recognized us from the workshops.

"He seemed to always be working on his house. Either by himself or with Ezra. He would collect usable housing parts after storms, and until used, they were scattered about his lot. This appearance of a perpetual construction site was not well received by my parents and others, according to my friends who lived near him. But on the positive side, he would always help neighbors with their construction needs, and many gave him construction materials they no longer needed.

"His home was damaged many times, and at some point he had flood insurance. I know this because the City had identified his property as a repetitive loss. 'Repetitive loss' is a FEMA term that refers to property with repetitive flood insurance claims and potential inclusion in a program to purchase the property by the community.[1] Purchase cannot happen unless you have a FEMA flood policy. He was made an offer, but he didn't accept. He later told me that he was sure this was a gimmick for the city to profit from his land.

"After all, he thought climate change was a hoax, Allison emphasizes. "He shared this with me on each of the few occasions we talked."

"Shortly after this offer was made," L^2 continued. "Another storm made landfall. The storm surge washed under his property causing severe erosion, making his property an island during ordinary high water. A debris line formed on the rear wall of his house. This more than met the city threshold of being inundated more than twice a month, but somehow, he moved the beach back with a rented backhoe, and the city didn't act. He was made a second offer but again refused."

"Several years later, a relatively small storm washed away more land, and then there was no way his property wouldn't meet the city's twice-a-month inundation threshold. Building permits and occupancy permits were denied. He claimed that his property was being taken without compensation, but his lawyer friends said that he

[1] Repetitive Loss properties are those that have proven vulnerable to recurring flooding due to multiple NFIP losses. Over 195,000 repetitive loss properties in the United States have received claim payments of over $12 billion, which accounts for almost one-fourth of all NFIP payments since 1978. The CRS program defines a Repetitive Loss Property (RL) as "a property for which two or more NFIP losses of at least $1,000 each have been paid within any 10-year rolling period since 1978 (e.g., two claims during the periods 1978–1987, 1979–1988, etc.)." Congress defined a new subset of RL properties called "Severe Repetitive Loss Properties" to focus available resources on mitigating these properties when it passed the Flood Insurance Reform Act of 2004. Severe Repetitive Loss Property (SRL), as defined in the Flood Insurance Reform Act of 2004, is "those 1–4 family properties (and non-residential buildings) that have had four or more claims of more than $5,000 or two or more claims that cumulatively exceed the building's value."

still could reasonably use his property considering the risk. He just could not build a home on it. A mobile home, an RV, a "small house," or other livable structure that could be readily removed in the event of threat, but nothing permanently anchored to the land."

"I overheard him telling a friend that his lawyer friend told him that if he were granted a permit to rebuild and occupy his house, he could sue the city for compensation when the house was again damaged. Since they allowed him to build and live in an unsafe location." Allison said.

"Unfortunately, he had let his flood insurance policy lapse." L^2 continued. "Fortunately, I guess, the storm destroying his home was declared a presidential disaster, and he did receive several tens of thousands of dollars through the FEMA Individual Assistance program and some FEMA temporary housing assistance. He eventually recognized that these funds would not make him whole, and it was about this time he went to live with Ezra."

"The CRC outreach worker, actually a student intern from Howard University," Allison said, "contacted him and told him that there was interest from a developer In buying his development rights. Density entitlements he could no longer use to rebuild at his beachfront site. He took advantage of the offer.

"I know Gadsden at least once thought about the possibility of applying some of the proceeds from the sale of these entitlements to purchase a condominium unit in Hilltop. It never happened. I do not think that he ever saw himself living in Hilltop or any multifamily building for that matter. But others did."

"Is this where the Purchase and Transfer of Development Entitlements or Rights (PDR/TDR) provisions the city adopted in their land development regulations come in?" I asked. "Maybe not for him, but aren't there many ways to configure a program in terms of entitlement?"

"Here, the TDR approach was pretty straightforward." L^2 continued. "The CRC created receiving areas in the lower-risk areas of Hilltop. The higher risk areas along the Troy Beach shore were the sending areas. The Hilltop receiving areas were upzoned to allow greater housing unit densities and commercial intensities, but only if the developer in the receiving area acquired housing unit or commercial square feet entitlements transferred from the higher-risk sending areas of Troy Beach to existing Hilltop properties. Developers, mostly in partnerships with existing Hilltop property owners, purchased development entitlements from the Troy Beach sending area property owners."

Ozell came over. "Have you had any refreshments? We have quite the spread."

"Ozell, good you're here," I said. "You could clear up a point. Why did you agree to share in the profits? You approved a zoning change to allow for greater housing density. Then you gave some of that profit away to purchase development entitlements. You would have achieved your goal of increasing the number of housing units in Hilltop without having to purchase entitlements from Troy Beach.

"Part of a compromise," Ozell said. "We were willing to share some financial gain with Beachers, but there was a limit. Buying entitlements transferred potential profit from Hilltop to those in the higher-risk sending areas. We wanted to encourage them to relocate here. But we also wanted to see a financial gain for Hilltop property owners. Without it, we may have pursued development independently. Developed without including Troy Beach in the discussion. You can think of the TDR process as marketing."

Ozell smiled while leaving to return to the other celebrants.

"The CRC also prepared sub-area plans for both sending and receiving areas," L^2 said, "to favor development needed to advance commercial, energy, housing, and social objectives. For instance, in Hilltop, where rising seas were not a concern, adapting to increasing temperatures and severe winds was. Strategic planning required appropriate urban design.

"Also, the sending area's sub-area plans in Troy Beach dealt with providing an acceptable orderly transition. This is what Paige and Cap were referring to when they mentioned that interim protective measures were built in Troy Beach to buy time, as they referred to the effort.

"These plans were begun way before damaging thresholds were crossed in Troy Beach. Part of the IF / WHEN contingency planning effort".

"I guess the price of entitlements depends on the market demand," I said.

"For both sending and receiving areas." L^2 agreed. "Densities in the sending areas could only decrease. Accordingly, at first, the demand for these entitlements was low. Until that is, the CRC formed a bank and offered to buy entitlements at an economically established market rate. Prices jumped substantially even before the plans were made public. The majority of available entitlements were sold well before 'sending area' homeowners crossed any thresholds. A provision was made in the program that enabled entitlements to be transferred with life estate remainders. This meant that the selling homeowner could remain living in their home until the owner died, damages to the home crossed a regulatory threshold, or the home was destroyed by a storm or other disaster."

"I have heard that 'density bonuses' are being made available," Watson said. "Bonuses for adding open space, reducing unit costs, and supporting infrastructure needs. Don't these reduce the profit?"

"They add community benefits," L^2 said. "But if they are not profitable the developers will not use them. All I can say is that they are being used. And we are achieving our objectives.

"Watson, you also need to include in your document, that the 'Thatcher Group,' and other developers, are partnering with the Hilltop landowners. The owners—heeding Ozell's son, the lawyer's—advice, seldom sell outright. Most want to maintain an interest in their community. Some leased the land to the developer. Some became partners in the development and are provided with a profit percentage. This profit has been received as dollars, but in many cases, it was in housing units or commercial area entitlements.

"Speaking of what you should include in your book. Don't forget that the CRC initially offered Gadsden a life estate. Life estates are also helping to strengthen the purchase and transfer of development rights."

"Let me try to describe this," Allison said. "I had a real estate course at State. Property Rights, can be described as a bundle of sticks. Entitlements to the property owner. In Troy Beach, if you own property in a residential zone and build a home you can paint it any color, have a pet. Sell all or some of your rights. If you are going to paint your house your University's colors for instance, say black and gold for the University of Iowa, and your neighbor hates the university you could sell that right to your neighbor and it can run with the deed. The existing and future owners will not be able to paint the house gold and black. This is a minor example."

"In Gadsden's case the CRC offered to buy his right to transfer his property entitlement units upon his death," L^2 said. "They also offered to purchase all property entitlements should the home be substantially destroyed for any reason. With the rapid rate of erosion and the water marching landward, this offer was probably not in the best interests of the city. A moot point. He didn't accept anyway."

***Purchase and Leasing Back—Atlas*:**

"You should include what Atlas did," Allison said. "After a storm, there was extensive damage to a block of homes in back of Front Street. The FEMA Risk Maps program in cooperation with the city had prepared future condition maps that illustrated inundation rates. At the time, FEMA referred to these as Graduated Maps, and they illustrated areas that had ten-year, twenty-year, and thirty-year estimates of being normally dry land. That is, before the land would be under water twice a month. The higher frequency inundation areas were designated sending areas, and the City was able to take ownership by purchasing development entitlements. Some of the purchase funds came as a result of FEMA pre-disaster and resilient communities and infrastructure grants."

"Leases were then sold. Atlas bought most. He is building the modular development you saw from the old oak tree," Sherl said as he joined the group.

"A lot of the labor was provided by community college graduates. These graduates are in high demand. Many communities are building similar developments."

Flood Insurance, Hazards Mitigation, and Comprehensive Plans

"What should I include about the National Flood Insurance Program?" I asked.

"You probably have been exposed to all that needs to be said," L^2 said. "The NFIP worked well for Ezra and the Valley Subdivision. The maps will always be off, but that doesn't seem to bother the residents. The regulations reflect the flood elevation estimates on the FEMA Flood Insurance Rate Maps (FIRM). Flood insurance is available for those who can afford it, and for others, its availability helps keep the housing affordable. So, all is working in Valley."

"For Troy Beach proper, not so much," Sherl interjected. "In unsafe areas where the hazard is not just flooding, but erosion and the gradual non-retreating inundation caused by rising seas, the mapping of probabilistic flooding is not helpful. It allowed users to think that traditional mitigation, elevating, and hardening or otherwise strengthening their homes, works. It's only temporary at best. Insurance is predicated on the wrong risk.

"FEMA is considering a Coastal NFIP, or CNFIP, where the hazards are illustrated in erosion/inundation rates, as reflected on graduated maps," I said. "But you've already discussed this."

Hazards Mitigation and Comprehensive Plans:

"Do you have any thoughts on what should be included about the Hazards Mitigation Plans themselves or the process used?" I asked, looking at L^2. "Were they helpful? Were the comprehensive plans supportive?"

"Same problems as with the NFIP. They achieved their objectives when addressing the correct hazard. But not when they didn't." L^2 said.

"At the time of Hurricane Apollo, all plans supported each other," L^2 continued to explain. "The Hazards Mitigation Plan, or HMP and our Capital Improvement Plan or CIP. They all relied on the FEMA FIRM and promoted stronger development codes recognizing the illustrated flooding risk. Nature-based solutions were also part of the HMP. These included planting dune grasses and building protective elevated boardwalks over them for people to access the beach without foot traffic causing erosion of the dune plantings. The HMP recommended protective hurricane-safe refuges and training for vacationers. None of these recommendations addressed the sea's capture of the land."

You need to stress that, Watson, in your writing, Allison inserted. "That you cannot think of the sea's march landward as *flooding* but as *lava*. Lava doesn't retreat. And neither does inundation. And a very small amount of vertical elevation of the water level can cause a remarkable amount of horizontal inundation."

"The Comprehensive Plan at that time called for high-rise multiple-family development along the shore, exploiting the beach, the dunes, and the view," L^2 said. "The towers were to be built to the newer, stronger building codes and were to be protected by more stable vegetated sand foredunes. The CIP had allocated funds to retrofit publicly owned infrastructure. To protect the facilities in place from hurricane winds and storm surges. Funds were included for dune work. The city's operations budget had a category for hurricane awareness training and outreach. All plans and budgeting documents were in sync. All embraced a similar trajectory," L^2 concluded. "Occupy with more human housing and harden in place."

"None of these documents pointed to the abandonment of emerging high flood-risk and future inundatation areas," Allison said. "These plans were driven by FEMA coastal flood maps prepared years before, and based on data sets collected years before that. Troy Beach was being made safe from threats that historically *were*. Not what was *expected* and eventually inevitable."

"Can I say that Troy Beach was being made a safer 1960s community, not one reflecting today's threats?" I asked.

"What you could say in your book," L^2 said. "is that Troy Beach was working under an aspirational comprehensive plan for a future community vision that could never be realized. Providing protection from the wrong hazards, and funding the hardening of infrastructure that did more to reduce safety than reduce long-term risks."

"'Disasters by Design,'" I said, "as an old colleague, Dennis Mileti, would say."

"We really shouldn't be so hard on the NFIP," Sherl said. "It is working for the Valley Subdivision, and it worked for Troy Beach while the threat was occasional flooding. When dry land was the normal condition. And the threat of flooding was restricted to that atop perceptively stable sea levels."

"I remember that after one storm a house fronting the beach was made into a fortress. The owners worked with energetic FEMA technical staff, and the resulting home actually withstood a direct hit from a tornado in a subsequent storm." Allison said.

"Pictures of their house appeared in newspapers, in online news apps, on TV news broadcasts, in FEMA publications, and in peer-reviewed journals," I said.

"We thought, at the time, that was a good thing," L^2 said. "Right? But the house proved to be an annoyance. Even eventually absent any habitability sustaining infrastructure for power and water, and the city's abandonment and vacation of the public interest in the street, it was one of the last to be razed. Neighboring structures were taken by the sea or removed following a storm. That building remained a testament to an old paradigm. The owners--a lawyer and an engineer--fought the transition with all the power they could muster. They met with their elected officials, friends of elected officials, and friends at the Army Corp of Engineers. Tried to have their beach renourished and armored. High tides washed over their property weekly before the owners accepted their fate."

"The owners were a lot like Gadsden, but with money and political connections," Allison said.

"I remember this story," I said. "The city was put in the awkward position of fighting its own residents. Residents who had followed the rules and had done their best to be good neighbors.

"'You can't cross the same river twice.'[2] I've used this in my class and asked why. Of course, this is because you cannot stay the same. Nor can the river. Both are only the same for a moment in time. Yet, these plans you mention often assume stability for natural and human-made environments. Some mitigation plans use future hazard conditions to define risk but then apply the analysis assuming that that impact will remain unchanged. Similarly, a comprehensive plan is driven by a community's aspirations, but their vision might not be achievable tomorrow. Capital Improvement Plans are often even more anchored in the present. CIPs may schedule maintenance or future work that preserves infrastructure that may be no longer needed or could even be harmful. The home fortress you described is a case in point."

"And it didn't help that at the time of Apollo, FEMA allowed the community to define the hazards of concern," L^2 said.

Transitioning to Hilltop:

"What about the transition to Hilltop?" I asked.

"Hilltop's acceptance of Beachers demanded a new community design, different street pattern, larger waste system capacities, and increased ability to manage stormwater," L^2 explained. "More parks, schools, and commercial uses. As Troy Beach lost population and infrastructure, Hilltop had to correspondingly gain

[2] RavenGeoffrey, Kirk, et al., The Presocratic Philosophers, Cambridge University Press, 1984.

services. Think of an old waterbed. When you push down on one side the other raises."

"Waterbed?" asked Allison.

Ignoring Allison, L^2 continued. "Here is where slow response was a benefit. We were provided time to play around with alternative scenarios. The IF / WHEN, THEN, THEN, approach evolved into a true contingency plan with an abundance of unknowns or at least uncontrollable opportunities and constraints. As hazard thresholds were crossed and property owners were restrained from new building and rebuilding, alternative approaches emerged. Some tipping points would be crossed slowly, others abruptly. The city would be pursuing a plan, and then a storm would arrive, the need for new directions would surface, and disaster recovery funding would appear, bringing a host of new challenges along with additional opportunities.

"A lot of credit for relieving this pressure to return to an old normal from the past, to fight nature, needs to go to Richard," L^2 said. "He was initially our FEMA Federal Coordinating Officer for several of the more recent disaster declarations but is now the FEMA Regional Administrator. He helped assure residents that Hilltop could become a feasible future relocation site for displaced residents.

"FEMA supported this transition to Hilltop with their support of the city's continual requests for Alternative Project funding. The first example was Front Street. Troy Beach determined that the public welfare would not be best served by repairing the Front Street facility or its function to its pre-disaster condition. I helped complete the paperwork and I remember the rule, word for word. 'A project was eligible for 90% of the Federal share of the costs that would be associated with repairing the damaged facility to its pre-disaster condition, or the actual costs of completing the alternate project, whichever is less.' Funds were diverted to Hilltop, as an Alternative Project.

"The CRC had prepared a local sub-area plan and had passed a bond issue to fund implementation, so our local funding share was available".

"Bond issue?" I asked. "I'm sure you are referring to Tax Increment Financing. Should I discuss that in the book?"

"Probably not." Allison offered. "Planners are familiar with this form of public financing. It has support in some circles and disdain in others. I am not sure if it is even legal in all states. Just mention that it was used to finance public infrastructure and the bonds are being paid off through resulting increases in tax revenue. That should be enough."

L^2 continued, "The relocation of our city hall to Hilltop, as mentioned by Angela, is an example of a Public Assistance Alternative Project that should be highlighted in your book.

"Richard also helped us apply for FEMA funds to remove abandoned buildings as they became unsafe either because of a loss in structural integrity or because the damaged building could not be legally repaired under our revised zoning code. Remember that our code was revised to prohibit the occupation of buildings that were partially or wholly over water. 'Over water' was defined by your two times-a-month threshold.

"Before we leave the subjects of thresholds, were there any other thresholds considered that needs to be mentioned? That could help other communities?" I asked, looking at L^2.

"Many," L^2 answered. "The future Ordinary High-Water mark was one. Just follow the shore debris or 'wrack' line. Or the Mean Higher High Water line. Someone suggested the reach of an average storm surge. Another was the FEMA 100-year or even the 500-year floodplain. All of these were discussed with the community but were determined to be more difficult to explain. King Tides occurred often and were visual and easy to explain. This might not be true for other communities."

Resilience:

"Well, I guess there's one concluding big question: Is Troy Beach more resilient?" I asked, looking at Allison.

"To begin with resilience is not a state, an end point in time," Allison began. "There is no assigned time horizon. Resilience implies having the capability to adjust to change when threatened. If we cannot, or choose not to reduce the threat, reduce carbon emissions, then we are given little choice but to change. Over and over again as the impact of a changing climate becomes more severe.

"There was always change, change is not new, but with our historical seemingly stable climate, change was occurring within a small margin. Metaphorically speaking, past generations might have had to walk to keep pace, but it feels like we are experiencing our lives on a skateboard. Our kids will need some sort of surfing hoverboard to survive.

"Is Troy Beach more resilient? The residents are adjusting to rising tides, and beginning to address rising temperatures. I do not know how they can't help but, in the process, accept 'change' as a way of life. As for our built environment, we are trying to adapt better. I feel Atlas and Thatcher are on the right track. As for the natural environment? Planting more trees and constructing more T.R.E.E.S. helps." Allison pauses to see if any got the pun. She saw a few comforting grins. "But, we must continually adapt to continuous change."

"Thanks for all your help," I said. "That's a lot of information. You've made me hungry. We should go off to the banquet."

Building Your Own Resilience Story

Falling Action: Approaches and tools, *clarified*. Challenge—Limited resources

The key tools employed by our planners are described. HOW will other communities enable their use? Action is beginning to fall as the context for "The Reveal," uncovered in the preceding chapter, is explained (Fig. 7.1).

In our story, Watson feels we need to provide some detail on the approaches and tools used. Depending on your intent, you might not feel a need to expand your basic

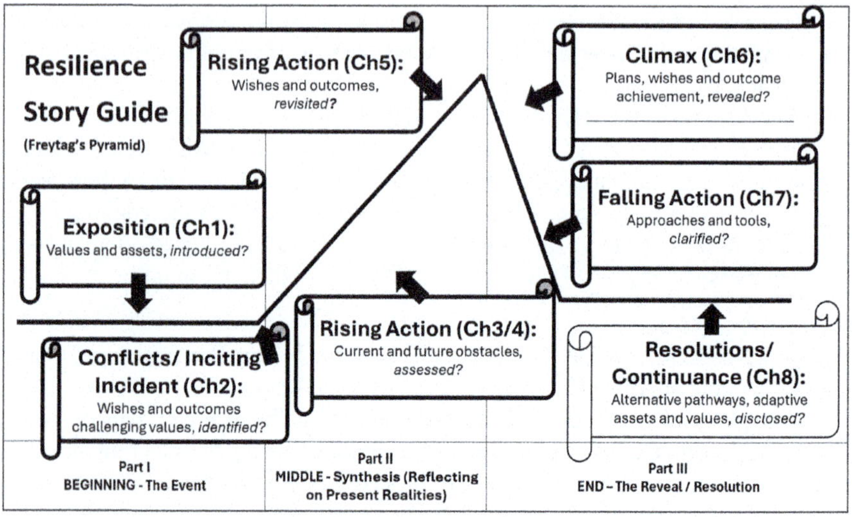

Fig. 7.1 Freytag's pyramid—action falls, approaches and tools are discussed. Prepared by Bob Freitag

story. However, we expect your story will be told and retold, and the process will eventually involve design professionals. So, this is for them.

Plot Development/Conflict Resolution/Resilient Conflicts

Your story reached its climax in the last chapter. It now exhibits falling action as details and context are provided. This does not mean the content here is any less important.

Falling Action:

Action falls after the climax, details are provided. Interest remains high.

Questions:

- What are the approaches and tools needed to bounce to that desired. Wished for?
- Can you offer the reader/listener insight on how they may apply your information?
- Are there any lingering tensions, unanswered or unresolved or unanswerable questions, or surprises the reader/listener should be aware of?

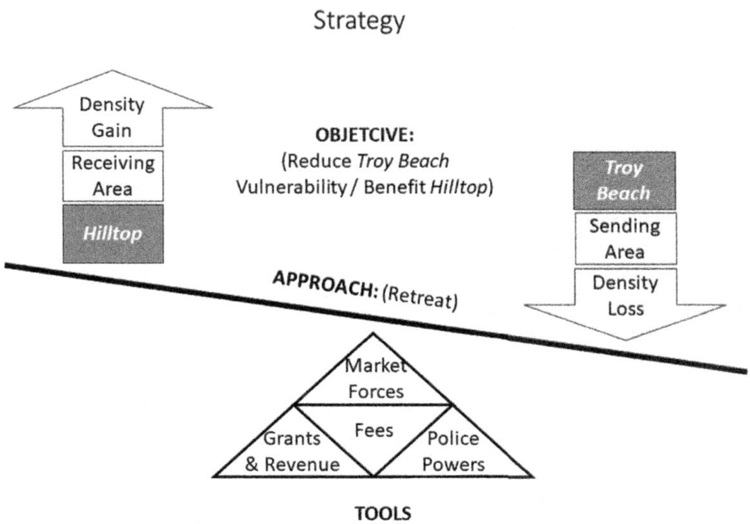

Fig. 7.2 Illustration of the tools used in support of an approach, where built capital densities are transferred from the Troy Beach sending area to the Hilltop receiving community. Prepared by Bob Freitag

We used a range of tools to implement our migration strategy.[3,4] We had to be careful of negative consequences. We created revenue by selling bonds to finance needed infrastructure, but we were careful not to end up with an overly indebted community. Market forces were exploited with the purchase and transfer of development entitlements, thereby incentivizing the relocation of habitable occupation from Troy Beach to Hilltop. Initially with a city-backed bank to assure purchases with willing sellers, but eventually entirely market-driven as Hilltop development and redevelopment occurred. We used police powers to develop the program and assigned density bonuses and fees to improve public infrastructure and direct the capital to Hilltop. We took advantage of FEMA grants to assist Troy Beach residents recover financially. Not necessarily made "whole," but helped.

Defining strategies as the approaches and tools needed to achieve an objective, Hilltop's objective was to maintain a functioning protective and supportive community and government on which the residents had come to rely.

For Beachers it was to have access to the Beach. This core value relied on having a sustainable viable coastal ecosystem.

As density was reduced in Troy Beach, corresponding availability had to be provided in Hilltop. See Fig. 7.2.

---

[3] Georgetown Law, Georgetown Climate Center, Managed Retreat Toolkit. https://www.georgetownclimate.org/adaptation/toolkits/managed-retreat-toolkit/introduction.html?full, Global Environ. Change 26, 152–158.

[4] Housing and Urban Development (HUD), Department. Climate Resilience Implementation Guide Community Driven Relocation, 2024.

As stated, the tools required to support a relocation objective included police powers, ways to produce revenue, grants and fees for services, and incentivizing market forces. One key tool was creating a method whereby development entitlements and rights from Troy Beach, could be bought and sold or otherwise transferred to Hilltop. However, caution was exercised so that the vulnerable migration of Troy Beach residents would be provided with incentives while not reducing proportional benefits to Hilltop. Hilltop needed to be awarded market gain, sufficient to offset the added expense. And the relatively flood-risk hazard-free environment of Hilltop, when compared to increasingly flood-risk hazard-prone Troy Beach, made bonus densities an economic incentive proving beneficial to both areas.

Other tools included charging fees for mobile home parks, purchasing and leasing back uses with accompanying restrictions, and selling bonds to develop new infrastructure supported by revenues for future higher-valued uses. We received grants from FEMA and other Federal agencies before and after disasters.[5]

FEMA allowed Public Assistance disaster funds to be applied to alternative projects. These were difficult to implement in practice because they required a host of assessments, including numerous benefit–cost analyses. The IF/WHEN, THEN, THEN contingency planning approach provided the time needed to undertake these analyses.

You will incorporate different tools in your story. If you are unsure, have your friendly design professionals take a crack at the story. I guarantee, as I am one of them, the challenge will be like sardines to your neighborhood feral cat.

Paige's Story

Paige's story now focuses on the establishment of a new historic district. She sees a clear path forward. The implementation of her wish depends on many. She realizes this and the need for partnering (Fig. 7.3).

She could have chosen other paths. This will be discussed in the next chapter.

[5] Housing and Urban Development (HUD). Department. Climate Resilience Implementation Guide: Community Driven Relocation, 2024.

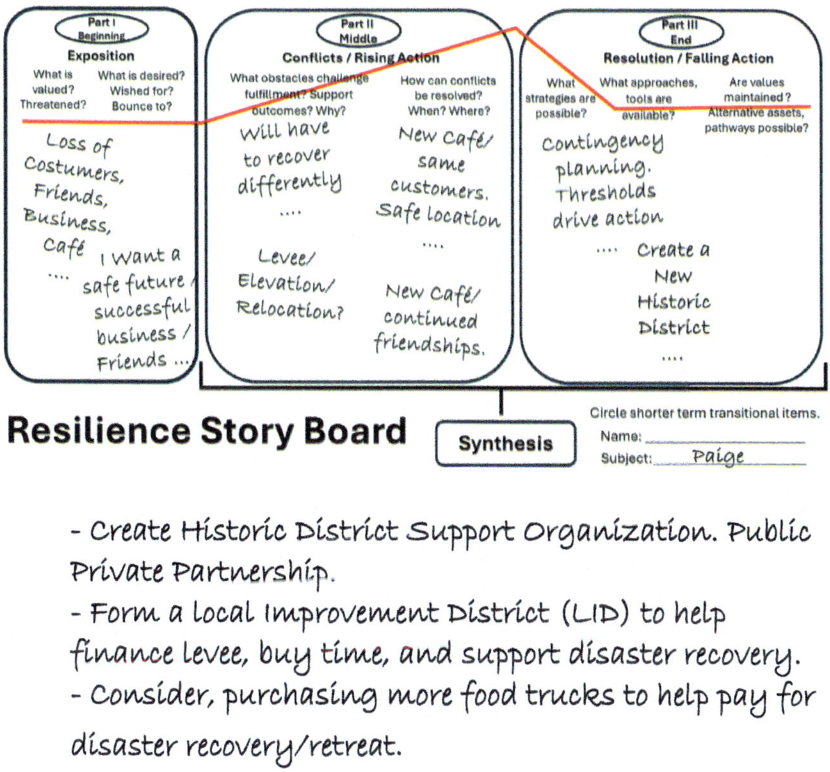

Fig. 7.3 Paige reveals additional adaptation strategies. Prepared by Bob Freitag

Chapter 8
Paths Not Taken

Alternative pathways, adaptive assets and values, *disclosed*. Challenge—HOW will competing voices be heard?

Abstract Before leaving, Watson goes to the more informal celebration in the Church basement and learns of pathways that were not taken but could have been. Here, additional context is provided. Troy Beach participants present alternative courses of action not followed, and why. Reader/storytellers are asked to assess their stories and ask if other courses may better support their values, and if there are interesting loose ends.

Keywords Alternative pathways · Scenarios

New Troy Beach Setting, Church Basement—25 Years After Hurricane Apollo

This collage of images has been assembled with the permission of the artist. Artist: Ron Kasprisin.

Resilience for Whom?

B. Freitag, *Resilient Shores: Adapting to Climate Change Through Storytelling*,
https://doi.org/10.1007/978-3-032-09047-8_8

When I reached the church basement, I looked over the gathering crowd to see who I remembered. The room was large, mirroring the footprint of the sanctuary above. Folks were arriving quickly and going to the snack table. I helped myself to a piece of mango pie.

Angela had just walked in and was, for a rare moment, by herself. I quickly took the opportunity to speak to her alone.

"Watson," she said, while leaning forward to offer a welcoming hug, but stopping in mid-flight when she saw the pie. "I hear you're writing a book, documenting our efforts. You should include a discussion of the paths not taken."

Ozell approached us and asked if he could join our conversation. I had not noticed before, but his neatly trimmed beard was now almost entirely gray.

Angela told Ozell, "I'm proud of what we're celebrating today. We might have taken a different path. Everybody made compromises along the way. Things could have turned out very differently. Better or worse, depending on your point of view, I suppose. Our dear friend Watson here needs to know there were other alternatives," Angela said, turning in Ozell's direction.

"From your lips to His ears." Ozell pointed straight up, and I prepared myself for some meaningful dialogue.

"Watson, why don't you and Sherl first go talk to Gadsden and Ezra while I circulate a bit? That's my job you know. Then we'll talk more about how we got here," Angela suggested. "Eat that pie you've got and talk to some of your other old friends."

I was a little disappointed, but I thanked Angela and Ozell. I located Gadsden and Ezra talking to each other in a corner of the basement.

The Beach Community—Stands Their Ground:

"It all didn't have to turn out this way," Gadsden began upon seeing me approach. Sherl walked across the room and joined us.

He began to describe a Troy Beach future he would have preferred. "If only the City had allowed us to continue as we had been. Respected our property rights. We would be better off. We all came to accept that rising seas and storm surges were causing the shore to move inland. Nevertheless, we could have adapted to the change. Differently.

"Sure, my house and that of my neighbors were becoming islands. Yet, we didn't need to move. We could have elevated them, and when access to our 'island' homes became difficult, we could have built boardwalks to the shore. As seas grew higher, floatation could have been placed under our homes, allowing our houses, our private property. to float above the rising seas.

"I understand that this strategy may have been a better alternative in protected bays and harbors, but it's used in other parts of the world. There's no reason why breakwater barriers couldn't have been built offshore—to reduce destructive wave forces".

"The city could not guarantee that the breakwater would protect leeward development in the structure's shadow as the number of stronger storms increased," Sherl interjected. "The approach would easily cost many times the value of what they are protecting.

"Also, as the shore migrated under your house, then continued shoreward, public access to the beach would be prevented. Your beach home would form a wall in front of the once-public beach. Access to the beach was valued. That was the main reason most moved here."

"Not my problem," Gadsden said. "American values revolve around private ownership. I got access to a sandy beach, and I should have been able to enjoy it. If others were equally industrious, they too could have enjoyed the privilege. But their laziness or lack of insight placed them in a less favorable position and this misfortune should be the government's responsibility.

"As for the cost of protection. Protecting us is THE role of government. And the protected area could easily generate additional upscale development. More density, more tax revenue for the city. This would make the relative value of the breakwater more favorable."

Gadsden inherited his property, and he probably would not have been able to pay the higher taxes associated with the higher use. I didn't feel this would alter his conviction and I kept it to myself. I felt that more density just meant more at risk.

"Watson," Ezra interrupted, ignoring Watson and Sherl's comment. "There are great advantages to what Gadsden is saying. With structures extending out into the ocean, and with the shore eroding, our sandy beaches would slowly disappear anyway. So slowly at first that many may hardly notice. This gradual shoreward progression of the beach may be seen as bad, but to Gadsden and others like him, the more resilient among us, our sandy beaches, may be viewed as a coin with both negative and positive sides.

"Access to the sandy beach is often seen as beneficial, but the fact is that beaches, being an attractive nuisance, often bring undesirable visitors who can cause a disturbance. As the beach disappears, we can be more protective of our property rights and we will gain unencumbered access to the ocean. Our elevated, then floating communities will have unobstructed ocean views—a protected, safe place to moor our boats and live an unfettered life."

I glanced at Gadsden and then at Ezra, unable to think of a reply. I had heard all of this before, as had Cassandra and the planning team so many years ago. Our job as planners was to address the values of the larger community and not merely coastal property owners. We tried to work with nature to come up with the lowest-cost, most-effective solutions. His ideas did not resonate with the larger community.

"Here are some more thoughts." Ezra continued, speaking more formally as if rehearsed. "But before I begin, I must tell you that these are not entirely our ideas. It reflects my favorite author Ayn Rand. You have to realize that Gadsden and I spent many hours refining and honing our arguments before we presented them to your planning team, back in the day when the city's direction for Troy Beach was being established. The city rejected them. Troy Beach is not the only community along the shore having these concerns. We have been talking to many other communities over these past years. Our arguments have not changed. They are resonating with many.

"Let me begin by saying that we needed to build better. All agree with this. Beachfront homes will have to accommodate rising seas and storm surge increases. This will take capital, 'hardening' homes can be expensive. This restricts ownership to the more resistant among us. Those who cannot afford to keep pace with a changing shore and are a liability to the community would be forced to sell to buyers who have the means to be safe. New building codes could be adopted, but it would be better if we could repair the homes as we feel needed. There would be less risk and less damage following severe events. Less public expenditures. Less drain on our collective government resources. The individual efforts of the more creative among us would provide a beacon to others. However, there are negative sides. Liabilities. With less damage, there will be less opportunity for those with the means to assemble properties and amass developable capital.

"You may think it strange that gives tacit support for stronger codes. Well, I have been slammed with floating housing debris washed into my house and causing considerable damage. Some requirements may be needed. But codes should not detract from individual creativity.

"We also agree that we need to increase affordable housing. We Americans are terrible at building new affordable housing. But there are two ways to lower the market price of existing homes. Removing amenities is one and what better surefire way is to limit sandy beach access? A wall of gated elevated floating homes blocking access to an already eroding beach is a guaranteed way to increase the inventory of low and moderate-income housing."

"Troy Beach could have been an example for others," Gadsden said. "These elevated and floating communities would have demonstrated what success looks like, thereby providing an incentive—an inspiration for average Americans. Attractive floating home communities will support our basic American value of independence. And if these homes increase in value, it is just more a sign of success, of a life well lived. Material and financial success can be a sign of divine favor."

"We would be preserving property rights," Gadsden said while stepping more toward the center of the group. "What right do others have in threatening our privately owned platted lands? As seas migrate inland, they may cover our lawful claims, but not dissolve them. Our continuance of ownership cannot, rising seas or not, cannot be challenged."

"Gadsden," I said softly, carefully choosing my word. "How can you say you're some kind of 'Fountainhead' architect? You had little money, and after Hurricane Apollo, you no longer had a livable home on the shore. You had to depend on a whole slew of people and government programs to help you rebuild. Now you are promoting a more restrictive community. One that you most likely couldn't join. I do not understand."

"I worked hard for what I have, and I do not want it taken away," Gadsden said. "If I could have kept my property and continued to redevelop, was allowed to sell to developers I would have been a rich man. I would have worked day and night to make it so."

"We feel," Ezra said. "That what we are describing would be a more natural, less intrusive, organic, green-designed, risk-reducing evolutionary process, than one of forced migration to Hilltop."

"Ezra," I said. "You live in the Valley Subdivision. Not even close to the beach you are suggesting restricting access to. And you could never have afforded to buy there, at least right next to the Troy Beach of old. I just do not understand why you are promoting a future in which you cannot access."

"I hear you are documenting what Troy Beach is doing," Ezra asked. "Be sure to include directions they could have taken."

I told them that I would do my best to document all sides. I then asked Ezra about what was happening with the Valley Subdivision.

The Valley Subdivision—Flooding Benefits:

"We still have to recover after every flood," Ezra said. "Floods are coming much more often. However, with each flood, we get more used to it, and the alternative for many in Valley would be to go homeless. The flood insurance regulations that FEMA still requires the county to enforce do not present much of a problem. After every flood, some homes are required to elevate and, in a few cases, the County offers financial assistance. Most adapt, however, little by little after each flood. They move furnaces, air-conditioners, and other appliances higher off the ground. The flooding is only a foot or so deep in most places. Like before. It just seems to happen all the time, now."

"There is still little water velocity in the river when it floods, and not much large debris floats down the river," Gadsden adds. "So, their problem is with stuff getting wet, and not the houses being washed away.

"Living better, with flooding, is the goal. Having no long-lasting disruption. Valley is achieving this goal, little by little, with each flood."

"For instance," Ezra said, "The basement of our Southern Baptist church used to flood a lot. It's a grand church, as I am sure you remember. You can see the steeple way up the valley and hear the bells as far down as Troy Beach. The basement flooded three times in one year. With our hotter weather, the mold set in. A real mess, and some of the ladies got sick while cleaning. We finally filled the basement with gravel, and that solved the flooding problem. This was after the last time it flooded. The flood was the result of a large hurricane, so FEMA became involved. FEMA gave the church some money to help pay for the gravel. The Church was our only place to meet as a community, so it functioned as our community hall, and FEMA was good with that. We need more space for the youth and other groups to meet, but we'll figure it out and build an assembly/gym next to the church. The structure was elevated."

"Don't County regulations still require most repairs to comply with a host of flood damage prevention codes?" I asked. "If the repair costs exceed half the value of structure, doesn't the entire building have to be brought up to code? That is if it is located in the FEMA floodplain."

"The operative word here is probably 'exceed,' and who determines that a structure has exceeded fifty percent," Ezra said. "You can do a lot of repairs that's under this threshold."

"Especially if you do the work yourself," Gadsden adds. "If you take a few years to complete it. Most in Valley are handy with their hands."

"Our Troy Beach High School biology teacher lives right alongside the river where there is some water velocity," said Allison who had been listening to us for a while. "The teacher planted bushes upstream of their house. These plants slowed the current, causing sediment to deposit among the branches. The process has created a nice vegetated and sustainable protective mound of dirt in front of their house. It diverts harmful flows and debris away from the house.

"What most of us do when we hear there is flooding is just turn off the lights, the gas line, for those few still using gas, and water, and move in with friends that live in safer homes. Some just leave town for the week".

"There seems to be a bit of an ethical dilemma there," I said. "The point of the National Flood Insurance Program (NFIP) is to get people out of the floodplain. Vertically or horizontally."

"Having a vulnerable, floodprone home is better than no home at all," Gadsden said. "It only appears to be a *bad* home every so often.

"Some county staff seem to have a problem with us living a floodprone life instead of building safe and healthy housing. As Ezra said, mold is our major problem".

"I think that's why Cassandra asked me to observe the planning process," Sherl said. "I was a riverine flooding specialist, and I believe she felt I needed exposure to all sides."

"Some of the floodproofing ideas you had twenty years ago," Ezra said, "and were rejected then, are being implemented now as I just mentioned. It was one thing when it only flooded every decade or two, but now it floods several times a year.

"Do not forget that most of Valley residents are working class," Gadsden said. "When you floodproof your home, you may receive less damage, but you also make it more attractive. You increase the value of the home. This is a good thing if you want to sell. Not so much if you have to pay more taxes."

"This is a key point that Gadsden is making," Ezra said. "As valley residents, we're afraid of our community becoming too attractive. If property values go up, we won't be able to afford to live there. And not all the Beachers what to relocate to Hilltop. Our Valley Subdivision might be their best alternative. We have some of the least expensive housing in the state. And we're close to the beach.

"You made this point during the workshops," Sherl said. "What you're saying is that complying with the NFIP, updating the FEMA flood maps, elevating homes, will all increase property values and, in the process, taxes."

"I feel we can maintain our community as long as home values don't change a lot," Ezra said. "I cannot say it strongly enough; complying with FEMA flood insurance requirements will increase land values. This might seem attractive at first glance, but most of us want to remain in the area. Higher taxes would bankrupt many of us even if a higher sales price would mean being able to buy a little more expensive home. I doubt these 'little more' expensive homes are even available.

"For most of us, getting a conventional loan is difficult, and for some, impossible if your house is in a designated FEMA Floodplain. Fortunately, for now, at least, most are not looking for loans. We also like our older FEMA Maps. Those that are in a FEMA floodplain have to buy flood insurance if they're getting conventional loans. It's expensive. However, on the positive side, this helps keep some buyers away, and that can be a good thing. Keeps housing affordable.

"Also, the Church has created a little revolving loan program. Sort of like a credit union, they've made available to parishioners. And don't forget, a seller can always help finance the sale of their home with a sales contract, and not go through a regulated bank. These tools help. But most homes just rotate between families and friends.

"The County implemented a 'Buy-Out' program a few years back." Ezra continued. "Several Valley homeowners located in a designated FEMA floodplain, and because their houses flooded a lot, the County classified them as repetitive loss structures and offered to buy them if they wanted to sell.

"Right after the flood, it all made sense, but it took several years for the county to make an offer. By that time all had repaired their homes with money from their Flood Insurance claims—most lost interest. The county ended up buying one home, owned by our Boy Scout troop leader. We hated to see him leave. When he left, the troop fell apart. Now, the lot is empty, and it has become an eyesore. We've asked the county for some money to make the lot into a little park, with some play equipment. Currently, it just collects trash. But the County does not like to give grants in floodplains."

"That Buy-Out program has ended," Sherl said. "Was deemed too expensive."

"We have had conversations with the county about incorporation," Ezra said. "But not for the reasons you make think. The main reason we wanted to incorporate was that it would allow us to drop out of the NFIP. Sure, we would get less disaster help from FEMA after a storm, but many felt that flood insurance was too costly anyway, and building or rebuilding to program requirements just made the homes more expensive."

"You know that might disqualify you from receiving other FEMA funding," I said.

"That's true," said Gadsden. "However, FEMA disaster funding mostly goes to repairing damaged publicly owned structures. That's good for the old Troy Beach and will most likely be the same for the new Troy Beach. They will have a lot of expensive assets: roads, a bridge, and utility systems. But here in Valley, we don't have those things. We seldom get much.

"The Valley is largely made up of the workers that the more privileged Troy Beach residents depended on for services. We're the cooks, the housekeepers, the mechanics, and the carpenters. We keep Troy Beach running during good weather and repair their assets after disasters. We need to have a place to live. They need us. Just think what would happen to New Troy Beach or the tourist economy, without us. Their progress, if you can call it that, would come to a halt.

"Also, FEMA and the NFIP are taking ownership of some homes that have been substantially damaged and in a high-risk area as defined on their graduated maps".

"So, flooding is not all bad," I said, more to myself than to the others. "It can tear riverbanks apart and destroy homes. But we also need to say that it can transport nutrients and commerce, depreciate home values, and maintain affordable housing for those who have no other choice."

Gadsden and Ezra walked away after making their arguments. Angela, Ozell, and Cassandra each found their way back to Sherl and me.

"See, it hasn't been easy," Ozell said. "Gadsden and Ezra aren't bad guys, but they haven't changed their minds in twenty years. Probably much longer."

Laissez faire:

Before I could respond, Thatcher came up to us. I hadn't seen the developer for decades, but she was still easily recognizable, dressed for high-stakes business with an eye to making a sale if the opportunity presented itself. "You know what we are celebrating today is not what would have occurred if we allowed the communities to recover naturally," Thatcher said, looking directly at Cassandra.

"Funny," I said to myself. "Gadsden and Ezra had made a similar reference. If only events were allowed to occur 'Naturally.' Only their definition of Naturally was very different."

"Hilltop would still have evolved from it being a sleepy, lost, little community to a growth area."

Ozell's eyes rolled back a little.

"You don't want to admit it, but you know it's true." She noticed Ozell and turned toward him. "Before your Troy Beach workshops, there used to be five hundred people living here, in Hilltop, on and among twice the number of larger lots. It wasn't even a town; it was a rural place by census standards and just a crossroads for most people. There were about five shops in what you called your Historic District, the downtown area, and that shabby old boarding house waiting to be upgraded to a decent vacation rental. The whole place was stagnant. It needed new life."

"The community had been a lot larger," added Angela. "But, over time, residents either died or moved away. There were always only a few jobs available in town, but there was always Troy Beach. It all worked. We had a lot of gardens, we grew our own food, and we always hired each other, and most importantly, we all looked after each other."

"It didn't work for your children," said Thatcher. "They all left for better employment. Yours included," she said, looking at Ozell.

"My son had expressed some interest in permanently returning," he said.

"After the city workshops, the interest in Hilltop skyrocketed," Thatcher said, glancing at Cassandra.

"Development could have leaped forward if just one owner had sold. Lots were large and workable. One owner, I remember, had a several-acre plot on Main Street and had a slight view of the ocean. If only I could have purchased that property."

"I know the person you approached," Ozell said. "His great-grandfather acquired the property during Reconstruction, like most of our ancestors. What's your point?" He stared at Thatcher.

"The point is," Angela interrupted, looking at Cassandra. "Before the city workshops, the place was worth under a hundred thousand. His son heard about the workshops and explored the out-of-state real estate market. He had an offer for well over five hundred thousand, and in retrospect, this would have been a steal."

"The son of the owner of the old hotel heard of this offer and almost convinced his dad to sell the property to you, Thatcher," Ozell said. "Rumor had it that the son was going to move back to take care of his dad, but that all changed when these offers started surfacing. He could have sold his dad's property for over a million. Sure, it included the beauty shop too, but still, a million dollars."

"Didn't know you knew that," said Thatcher.

"Then," said Paige, who had also joined the group, "Thatcher contacted me and asked if I could manage the property, the old barbershop, as a high-end coffee shop over the short term." She looked at the developer. "You were making other offers on the block, several of which had nice views."

"You knew that too?" asked Thatcher, surveying the group. "Okay, you also probably knew I'd been seeking a permit to build a condo next to what has become the new Hilltop Historic District. I was going to modernize the District, increase the density, and make an all-inclusive type resort."

"All exclusive, Thatcher?" Angela asked.

"You would have all been wealthy, Thatcher said. "I was offering a high price."

"Didn't you feel a little sad that most of the original residents would have to leave?" Allison, who had also joined the discussion, asked.

"Progress is progress," Thatcher answered, "and as I said, all of the owners would be pretty well off."

"Not too sure of that," Angela said. "Where would these residents have gone? They certainly would have lost their community, and what about our church?"

"I was going to preserve the building. Thatcher said. "It would have made a great tourist destination. I'm not too sure it could have continued functioning as a church. But that's okay. There was the Baptist church in Valley.

"If you all had not stopped me," Thatcher said, looking at Casandra, who was talking to others across the room, "all of it could have been built by now. Cassandra, if her team hadn't come up with your IF / WHEN strategy, and Angela, if you and Ozell hadn't convinced your residents not to sell. Well....

"We were so close. After the workshops, if I could have bought just one or two properties, most others would have sold, and with few of the original residents, I could have easily convinced the county to incorporate Hilltop as a town. The petition would have sailed through with very little opposition. Oh, well, I guess some people aren't interested in money." Thatcher retreated to the snack table and poured herself a glass of punch.

Hilltop Community—Stands Their Ground:

"Wow, you weren't kidding about other possibilities and other agendas," I said. "Since I wasn't in Troy Beach long after the workshops, how did you get here? Why isn't this a wealthy resort playground or some other 'profitable' outcome for the residents of Troy Beach and Hilltop?"

"Back when you were interning, we accepted that Troy Beach and Hilltop were under a lot of pressure to change," Ozell quickly began as if he had been waiting a long time for this opportunity to speak. To clear the air. "Even though we didn't know what was going to happen or when. Shortly after the workshops, our Hilltop community met together and worked with Troy Beach. We felt like we could trust them a little more, but we weren't part of the city. We were just in the county, so we were still wary. I felt better after the workshops, but even with Cassandra there, I wasn't buying into believing our Hilltop values and our supportive community would stay intact. Were we just going to end up with a new Troy Beach, on Hilltop land? At least that's what it sounded like to me at first." He shook his head.

"Angela and I knew Hilltop land was increasing in value. Just not as fast as it was actually going up. It wasn't the next day, but soon after the workshops, several out-of-state and offshore banks began making offers to buy Hilltop properties. At least one offer from an international developer."

"None of the developers said what they wanted to do with the land," Angela added "And they all had a similar approach. They made offers to clusters of property owners, at very good prices. 'Offers with no face', we branded them. We were a tight-knit community, so we all quickly learned what was happening. As Thatcher just found out."

"Fortunately, the Admiral's daughter was available," Angela said. "She had a PhD in planning, remember, and quickly realized the pickle we were in. Dr. Nguyen played the role of Cassandra, and we applied what we had learned from the original workshops. We met here in this church basement and held our own workshops. World Cafes and everything. She managed these workshops from the back of the room, so to speak. Just like Cassandra did."

Everyone glanced at Cassandra, but she was still engaged with other guests.

Angela continued, "The Doc, as we called Nguyen, let ideas emerge naturally, often slowly and after much discussion, without her direct input, just like Cassandra. We even invited some Tugaloo College, and State University students and faculty to conduct design studios. We had law students, budding architects, planners, engineers, and some pursuing public administration degrees."

"And a poet," Ozell interrupted. "I don't know how or why we had a poet. But the way he thought and spoke helped."

"The Doc, called the Workshops charrettes. I think that Cassandra might have also used the term but I cannot remember," Angela continued. "They built on what was experienced at Raj's. We also met in the Dirty Kitchen and discussed what we learned from our workshops. Our exercise produced a different vision for Hilltop. Students got course credit and experience; we got their help. Everybody was pleased. Their involvement brought a lot of energy to the process.

"The process resulted in us realizing our Hilltop value was not centered on the beach. What we most wanted was to preserve a community we felt safe living in. Building equity was important. Don't get me wrong. We all wanted to leave something for our children, and our grandchildren. But it was not everything."

"We had no control over the threat of climate change, of course," Ozell continued. "This part of the equation was certain. Our uncertainty was with being displaced by Beachers. How strong would the assault be? However, we still thought we had considerable control to manage, and direct the outcome. There was always the lingering fear of being forced out of our community by richer Troy Beach residents. They have few choices other than Hilltop.

"My son," Ozell announced with pride, toyed with the idea of us creating our own corporation, the Hilltop Preservation Corporation."

"The HIP," Angela added. "Cute name?"

"That's my boy, a law professor who's not so important that he forgot his hometown," Ozell said.

"For a while, he stopped coming back here. But after Apollo, he saw the reason, to preserve Hilltop."

"We also applied the IF / WHEN, THEN, THEN strategy that resulted from the Troy Beach process. Similar to the CRC, the HIP was to be a paper organization until thresholds were crossed. Unlike the CRC, which was a nonprofit, we were thinking of the HIP as a for-profit corporation."

Cassandra had now joined us, and we all exchanged welcoming smiles.

"Among the Hilltop folks," Angela continued, "we discussed a way to provide for the assimilation of Troy Beach migrants, the Beachers. We felt this approach would work as long as the pressure was not too great, and growth remained manageable. Over time, we would phase into a new entity and the HIP could assign stock credits to the contributing residents of Hilltop according to the value of their property or leasehold interests. These shares would only have material value when land was transferred to the Corporation, and this would only happen when thresholds were crossed. Our thresholds were levels of demand from the Beachers. We quantified these levels, but it is not vital to include them in your book, Watson.

"Our implementation tools were different from those discussed earlier in the sanctuary. For instance, with a HIP approach, we saw no benefit in having the Hilltop community designated as a development entitlement receiving area."

"Why should we give away value to Troy Beach?" Ozell interjected.

"With the HIP being a for-profit organization, we would be able to exercise considerable control over development. Protections for shareholders, leasing rates, and building design could be controlled by the HIP," Angela continued.

"Also, we didn't see an advantage in annexing Troy Breach. We did see an advantage to incorporating Hilltop as its own city, however, and we actually started the process. Creating a Housing and Land Trust was something else we began. We had some families doubling up who deserved their own homes. We also had affluent families who had left Hilltop, but were interested in returning. Becoming investors. We needed to help finance more affordable housing and expand our commercial area as Troy Beach's commercial area was reduced in size, and we felt we could capture a good piece of the market."

"We're still in contact with Tugaloo and State, and they helped brainstorm some ideas."

"It was critical to us that our Hilltop community not be exploited," Ozell said. "We planned on accepting dislocated Troy Beach climate refugees, but with conditions, more advantageous to our current Hilltop residents than originally suggested. We would accept Beachers, but it would have been on our terms. We wanted to create a process where we become better off for having Beachers move here and they wouldn't be worse off."

"Why didn't you pursue this?" I asked.

"We did," Ozell said looking at Cassandra for acknowledgment. "Many of our ideas were accepted. We just didn't need to create the for-profit HIP. My son became the CRC Executive Director for a while and helped draw up the articles of incorporation. Not while I was on the board, of course. As a result of the process, we all gained trust in each other."

"Opportunities to increase wealth that our older residents could pass down was a big motivation," Angela said. "We knew that our little community would change, but we also felt our children could be better off from us joining the Troy Beach effort.

"I felt we did not need a protective wall to shield us from a Troy Beach invasion. The HIP could have done that. For a while at least, but we all must remember that the Troy Beach community, along with developers, had a ton of resources. I think we could have won, but the CRC also gave us protection, and by this time in the process, we came to trust the larger community."

"Angela," Ozel interrupted. "I came to think of Hilltop as my great-great-grandmother Rose. She was incredibly beautiful according to our family story. Rose was separated from her mother when very young, as most enslaved children were, and she was raised by grandmothers."

"Okay, nothing really helpful here," Angela replied, trying to redirect the conversation. "We all have similar stories."

"As the story goes," Ozell continued in a softer tone looking at Angela, trying to persuade her that the story needed to be told. "The plantation owner took an interest in her, and this worried her grandmothers, her caretakers. As I said she was very pretty. She was sought after."

"I see a connection, but does this specifically help explain Hilltop today?" Angela asked, trying again to redirect. She felt strongly that this was not a story to be shared outside.

"There were not a lot of choices for her back then," Ozel continues undeterred and still looking at Angela. "She was bound to be exploited by the owner and when she got older, she would most likely become a toy to welcoming male plantation visitors.

"The grandmothers could scar her face and make her unattractive, but that would bring severe consequences. They could try to get her up north, but escape seldom worked. They saw only two paths. One was to get her into the Big House where the owner's wife might monitor her husband's actions, but that wouldn't help Rose from being a bed partner for guests. Also, there was a possibility that the wife couldn't control her husband and she would most likely take his transgressions out on Rose.

"What they decided to do, as the legend goes, was to get her into the house of the owner's son. He was a widower and had never beaten a slave as far as they knew. This arrangement might offer some protection."

"We all know the story," said Angela. "The son was Belle's great-grandfather.

"You're saying we can make Hilltop less attractive to developers, we could take the money and leave, or we take the best offer available and control the situation to the extent we can.

"And, Angela, these are not slave times," Ozell reminded her. "We have laws to protect us, and with my son and the Nguyen girl, we can exercise considerable control over our future. My son and she, along with Sherl, I think, created an organization to help us and other threatened communities. PAN, they call it. Stands for the Planning Action Network. It's associated with State somehow and maybe Howard and other universities. Not sure."

"Ozell, you make a good point," Angela said, "but the pressure for outside development would still be great.

"Casandra had something do to with creating PAN.

"And why do you think 'Valley' isn't targeted? Why is there so much development pressure placed on us?"

"Probably because those living in Valley were a little afraid, and they had the power to deflect initial assaults," answered Ozell. "Also, they are farther from the beach and don't have the views."

"Afraid of what?" Angela asked.

"Strangers," Ozell answered. "They have a pretty tight community too, with their church and all. I do not see them relishing an onslaught of non-believers. Also, they flood. A lot more recently."

"So, we accepted the increased interest as an opportunity," Angela said looking away from Ozell, at the others in the group.

"That's how I see it," said Ozell.

"Cassandra was always talking of common ground." Angela said turning to him. "What if the common ground of those owning land in Troy Beach was buying up Hilltop? Or, what if some small group in Troy Breach joined forces with outside financial and political interests and their uniting driver was in considering us prey? And they wanted to replace us."

"Well," Ozell said, "Cassandra and the CRC didn't let that happen. We always had the choice not to sell.

"Anyway, it didn't happen".

"Could have?" Said Angela.

"It did not happen!"

Ozell stepped forward, turning his head away from Angela, and looked directly at Watson. "In your book, Watson, please emphasize the importance of the workshops and the storytelling that was part of the process. They helped lay the foundation and the need for continued involvement. State clearly that the decisions demanded of us were / are really really difficult. And even that would be an understatement. You need to emphasize how the workshops helped cement trust."

Before I could acknowledge the request, Cassandra asked me. "Are you happy at Troy Beach's redirected path? What you are witnessing?"

"The more important question is are you happy with New Troy Beach?" I asked, turning to Angela and Ozell. "Is this the right direction?"

"Commonly shared values," both replied in unison.

Cassandra's Parting Advice:

"The answer will, of course, be different for everyone here, and it may take time," Cassandra said. "What I will say, is the future is now in the hands of those we have welcomed to this project. And want to thank you, all of you, for allowing me to participate in this journey. Please realize that even after your hard work, there is no 'There.' What you are creating around you will need to be changed. You will have to find and train those who follow you to take the reins, to learn, and to adapt, as they will have to teach those who follow them. Having the ability to re-organize only to re-organize again, has become our future. Resilience will rely on our children and grandchildren. And hopefully their children and grandchildren."

Angela had gone back up to her office to take a call. Now she came back into sight, walked halfway down the stairs, and clapped her hands loudly.

"Excuse me, everyone. I need everyone's attention." Being Angela, she got it immediately. "The National Hurricane Center has been following a tropical depression that has just become a Category One hurricane. They have projected possible tracks and we may expect landfall close to Troy Beach in a few days. You all know what to do."

This was a good time for me to leave. I said goodbye, but before I could retrieve my luggage Angela gave me a quick hug as I dodged that hat pin. Cassandra handed me an envelope.

I took an electric air cab from the new Troy Beach to the regional airport and caught my flight home. Still wearing my bucket beach hat, I took my seat and opened the letter. It read:

> Dear Watson, I want to make sure that the following key points are not forgotten in your book.
>
> 1. Values drive direction. Values define well-being. Hope, aspirations, goals. Our older summer occupants value visits from their grandchildren, Angela and Gadsden self-reliance, Paige friends, and Belle community. We had to work from values.
> 2. Values can be defined by assets. Values are lasting, less flexible, and often independent of attributable assets. Assets can often be replaced. Values not so much. Cap's asset was his store, other commercial efforts could prove alternatives. The beach was the community's key asset. Its value to Troy Beach was in providing views, recreation, destination, etc. Not in being right next to the shore.
> 3. Common ground. Common ground can be found in values, pursuits, assets, or fears. Gadsden, Atlas, Ezra, Ozell, and Thatcher could have fought for the control of Hilltop, or Valley. They could have picked a target, frozen it, personalized it, and polarized it. They didn't. They sought common ground and drew their circle of concern around all.

4. Assets are often masked. Assets include natural and social capital. Gadsden's acquired asset was his construction talent. Angela, Ozell, and Belle's talent was in creating community.
5. Hazards are constantly shifting, elusive, and unrestrainable. Hazards profiles include location, frequency, timing, and severity. Each of these characteristics is largely certain. Better to direct attention to their impacts. It took a while, but eventually all realized that the climate was changing and there was little, in the short range, they could do about it. Gadsden could not elevate high enough. Paige saw a sea wall as providing only temporary protection and knew that relocating would eventually be demanded.
6. Resilience may be more art than science. Strategies are approaches and tools that advance an objective. Their creation may be more an art than an algorithm. We had to thank Raj, and his creation of the Dirty Kitchen for this. The resolution of threatened values and assets is as much the result of art as anything.
7. Question past plans. Hazard mitigation and comprehensive plans may lead more to a better past or what can never be, then a better more probable future. FEMA maps and the community's reliance on them caused L^2 and Gilbert considerable distress. The folks in our community believed that they mirrored future conditions. They didn't.
8. Passion rules. Achievable outcomes require knowledgeable impassioned participants. We all have to thank Belle, Ozell, and even Ezra in his own way, for providing and maintaining the momentum for change.
9. Contingencies plans. Planning for the IF / WHEN, allowed for discussion that we did want to address. Creating thresholds help. The approach allowed Allison and Ezra to talk to each other. It kept Cap and Gadsden calm, and enabled Belle and Angela to do what they did best—allow everyone to feel good about each other.
10. Trust Governance. Troy Beach found solidarity with the creation of the Community Redevelopment Corporation. Its creation overcame some painful history between Troy Beach, Hilltop, and Valley. Troy Beach residents may have accepted their government being in charge. Hilltop wouldn't. Neither would Valley,
11. Resilience may have a darker side. Planning objectives must be recognized as just and fair. Present a planning process that is as inclusive as possible. Resilience is temporal and can be at the expense of other longer-term interests. At times we were walking a tightrope where we could slip into a battle with Valley, Tory Beach, and Hilltop. The resilience of Troy Beach at the expense of Hilltop or Valley.
12. Resilience is not an end state. You must present resilience as the ability to reorganize and continually adapt as needed. Paige and the Admiral caught on early that change would be a part of life.

Building Your Own Resilience Story

Resolution/Continuance: Alternative adaptive assets and values, *disclosed*. Challenge—HOW will competing voices be heard?

Resilience has no moral compass. Francisco Franco's goal was to achieve resilience for his concept of Spain, Reverend Martin Luther King Jr. was trying to achieve equity and ensure its resilience. How will your competing voices be addressed?

Our Freytag Pyramid indicates a fall, at this point, as the story of the HOW is wrapped up. The arc is complete, and the crises introduced have been resolved. Hopefully (Fig. 8.1).

Plot Development/Conflict Resolution/Resilience Concepts

I will end my assistance here, by returning to 5 Ws and H. This will be more of an assessment of WHAT IF, instead of IF / WHEN. What if you rewrote or retold your story, letting different characters and differing values drive your story? Or began with a different phase in emergency management or a different characterization of the hazard.

Thinking of our story as a community-based WOOP, what if blocks of community members had conflicting visions? How could common ground emerge? We did it by agreeing not to act until commonly agreed-to thresholds were crossed. Having built trust and creating a specifically tailored governance also helped. We also had Raj's where we had the opportunity to tell and retell their individual WOOPs. For us at least, a collective WOOPs consensus was spawned.

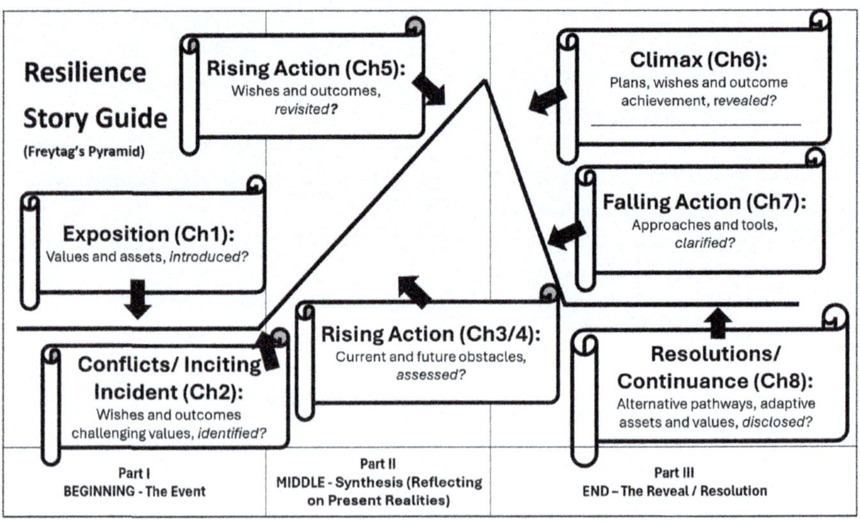

Fig. 8.1 Freytag's pyramid—alternative pathways. Prepared by Bob Freitag

Resolution/Continuance:
Your resolution concludes your story, well, sort of.

Questions:

- Have you completed your story arcs?
- Are any conflicts left open-ended? If so, have the issues been addressed?
- Are there any overarching themes or lessons that emerge that may need emphasis, to be helpful for the reader/listener?

Who: What if Watson was not the narrator? If the story was told by Gadsden, Paige, Ozell, or Allison? When you tell your story, dramatizing your community's future, try to tell it from different points of view. Or encourage your listeners, when they retell the story, to try to describe the plot from varying perspectives.

What: Watson's plot followed a re-organization resilience metaphors while tossing in some risk reduction hazards mitigation efforts and bouncing back to pre-event trajectory approaches to provide time. What if Watson's conclusion was to bounce back to an approach of hardening vulnerabilities? The story would be entirely different. Reconsider your resilience path or combinations.

Where: Our "Saving Troy Beach" is set in Troy Beach and Hilltop, with passing references to the Valley Subdivision. Watson had a class where students re-wrote the story from the perspective of the Valley Subdivision. The overall strategies were similar, only they applied to Valley. You or those you tell can have fun retelling your story. Differently.

When: We begin our story in the recovery phase of emergency management. It could have easily been written as an attempt to lessen the impact of an approaching event. In the preparedness phase. Or, although it would be more difficult to weave in longer-term goals, the story could have been driven by response. As a fire suppression mystery, the story could revolve around protecting Gadsden and his partygoers.

Why: The conflicts driving our *Resilient Shores* story were driven by the impacts of a changing climate to selected values held by our disaster archetypes. Watson could have built the conflict around other themes. Ezra's lack of acceptance of the science could have received greater emphasis, or Allison's questioning could have driven the storyline, as could an effort by Angel to protect Hilltop from Beachers. Beachers could fight Hilltop's unacceptance. Experiment with your stories.

How: Our HOW had its roots in contingency planning. Your HOW could be in providing good design with increased emphasis on adaptation or on increasing capital worth.

Our *Resilient Shores* story touched upon collapse but did not confront it head-on. You could.

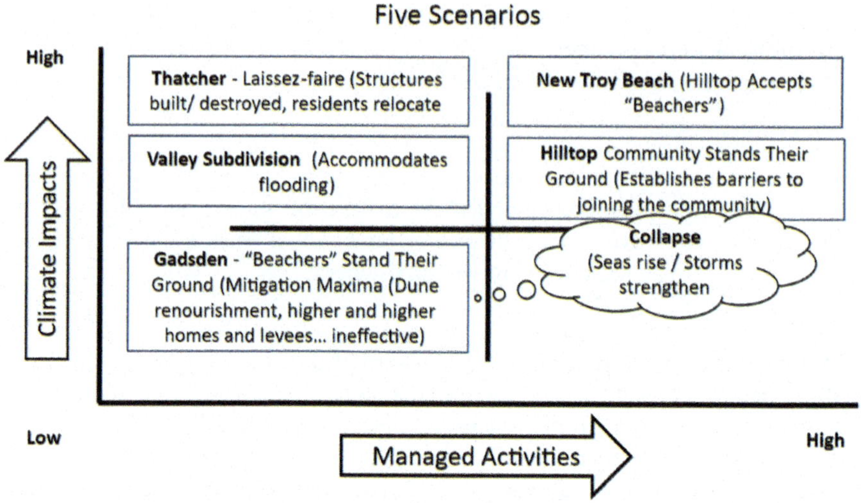

Fig. 8.2 The five alternative strategies presented in this chapter are illustrated as to their demanding high or low levels of management and being exposed to extreme and less extreme changes in the climate. The scenario offered by Gadsden is assumed to collapse as greater changes in the climate cause extensive erosion and home isolation, as the approach rejects alternative corrective relocation approaches

Scenario planning methodology can help when facing a range of possible futures. One approach illustrates future pathways as having two variables.[1] In Fig. 8.2 one axis illustrates what you cannot control (extrinsic variables), such as climate change. The other axis is designated for that for which you have control (intrinsic).

Gadsden described a future beachfront community that could accommodate a changing shoreline. For a while at least. This was a path that Troy Beach would have most likely pursued if we had not taken action. This was discarded by Troy Beach because it would cut off access to the beach thereby eliminating Beachers' principal value. We also felt it was not sustainable over the long term. The larger Troy Beach values would have collapsed if this trajectory had been advanced.

The Valley Subdivision resident's vision of their future was achievable. But this was not as a Troy Beach receiving area. Valley did not want to designate their homogeneous community as a receiving area for Beachers, even if flooding could have been accommodated.

Thatcher's alternative, which we called the market-based laissez-faire alternative, was attempted shortly after Hurricane Apollo. Progress (e.g. property purchases) was stopped by Ozell and Angela, but it could have easily forced an exodus of Hilltop residents. This alternative would have realized Hilltop's greatest fear. Hilltop would no longer support the community the older residents had enjoyed for so long. Hilltop, as defined by Angela, Ozell, The Admiral, and Dr. Nguyen, would have collapsed.

[1] Peterson, Garry D., Cumming, Graeme S., and Carpenter, Stephen. Scenario Planning: a Tool for Conservation in an Uncertain World. Conservation Biology, April 2000.

As Angela and Ozell explained there was a compromise in their acceptance of more common holistic values.

Paige's Story

Returning to your Resilience Story Board, here is how Paige might have completed this concluding chapter. Figure 8.3 She mused on what opportunities moving to Valley might have brought. You may now make an entry on your Resilience Story Board. It may not be your last entry however because you'll probably be altering your vision of the future with each telling—by you and each of your friends and colleagues.

Our story of HOW is complete, well, at least as of this writing.

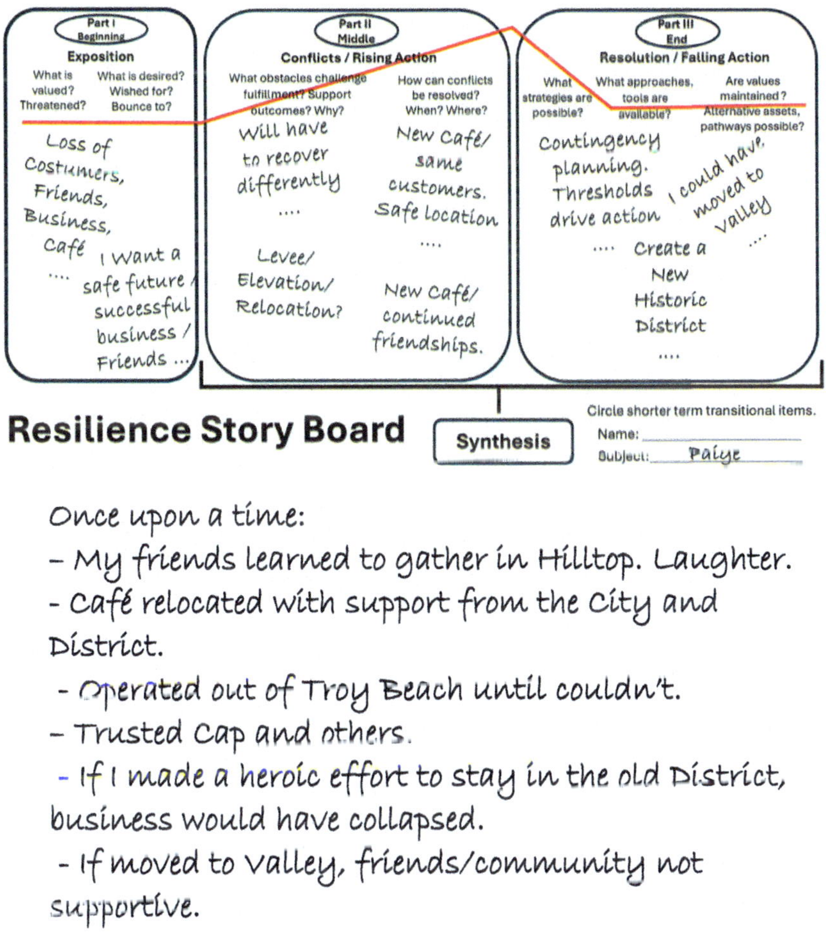

Fig. 8.3 Paige explores alternative scenarios. Prepared by Bob Freitag

What Next: Written by Cassandra

Watson has asked me to weigh in here.

We hope you enjoyed our story of Troy Beach. By now, you may have drafted your "Who done it?" or perhaps a "It won't be done again."—story. Your story has probably offered adaptation and climate mitigation strategies. It more than likely includes several implementation strategies and defines an overall trajectory for your subject of concern. Retreating or migrating inland may not be the preferred trajectory for your subject of concern, at least over the short term. Accommodation and protection may be favored. You may feel your target has sufficient trust in established institutions, and alternate forms of governance are not needed. Implementation can commence immediately. If you have crafted your own story, you have begun developing a plan.

In this chapter, I am asking you to revisit your draft strategies and view them as points of embarkation to corrective actions. Your strategies may have been influenced by the experiences of Ozell, Paige, Ezra, and maybe even me, Cassandra, or by your friends and colleagues.

Watson presented our Troy Beach and Hilltop family to help you visualize alternative futures for your target subjects, individually and in concert with friends and colleagues. But other paths could have been taken.

The American Planning Association,[1] at its most basic level, suggests that a public planning process include the Involvement of all segments of the community in:

1. analyzing issues, (outcomes and obstacles)
2. generating visions, (wishes)
3. developing plans, and
4. monitoring outcomes.

You were not asked to involve all segments of the community, develop thoroughly researched approaches or tools, or create a monitoring approach.

But you have started the process. Especially, if you invited others to share stories.

[1] American Planning Association, Godschalk, Rouch, "Sustaining Places: Best Practices for Comprehensive Plans, PAS Report 578, 2015.

B. Freitag, *Resilient Shores: Adapting to Climate Change Through Storytelling*,
https://doi.org/10.1007/978-3-032-09047-8

You have raised and analyzed issues related to the well-being of your subject(s) and generated future visions and suggested resolutions. Some resolutions were probably place-based, and some were not. You have either accepted, rejected, or altered, bouncing back as an alternative. You have drafted strategies and suggested trajectories as hypotheses to be researched.

In the following discussion, I will present guidance on possible alternative pursuits I found helpful.

Public Process

You should feel pride, deserving celebration for your story(ies). You are confronting climate change, and the need to adapt. You had to think about and undoubtedly discuss with your community, accommodation, protection, and even the possibility of migration inland with your friends and colleagues. Your stories probably allowed you to consider that that might have been almost impossible to consider otherwise.

Before you get too smug, you should realize that as good as your thoughts may be, they may be just your thoughts, and possibly those provided by your friends and colleagues. They reflect how you have defined your community, your perceptions of the risks, and your aspirations—concerns, hopes, and dreams. As you read and internalized the concerns of Troy Beach, Valley, and Hilltop, you probably could not have helped but devise a range of possible alternative strategies. Strategies that were designed to preserve the values presented. Maybe you even came up with competing values and stories.

You may have realized by now, that I, Cassandra, do not really have the personality or leadership abilities of a General/President Grant, Queen Elizabeth, Napoleon, the Great Khan, or Cleopatra. I could not have pushed forward the charges demanded of Troy Beach or Hilltop. Or, really, as good as I feel I may be, I probably shouldn't.

You, perhaps like me, are not a Frederick Law Olmsted, Daniel Burnham, or any of the others involved in the nineteenth-century City Beautiful Movement who were allowed to re-design entire communities. You are probably not of the same caliber as President Eisenhower or Francis Victor du Pont who created our Federal Interstate Highway system. Nor do you probably have the leadership abilities or convincing skills of Robert Moses who drove the renewal effort in New York, or the foresight of John Muir to help preserve our natural environment.

Of course, there are and were downsides to their leadership and power.

Many do not question the kings and queens of their professions. We must remember that even the efforts of these movers and shakers did not address the values of many, and precious assets were destroyed by them and their armies. The movements for which these leaders were responsible, were real threats, in and of themselves, to many fellow Americans. The resulting trajectories they advanced created real hardship for many, over generations. Inclusive processes are important.

To be able to celebrate what as revealed in Chapter Six, I had to craft a process whereby participants would internalize directions. They had to feel they were the creators. They may not believe me, but they would believe in themselves and their trusted friends and colleagues. Our public process became our strength, and it

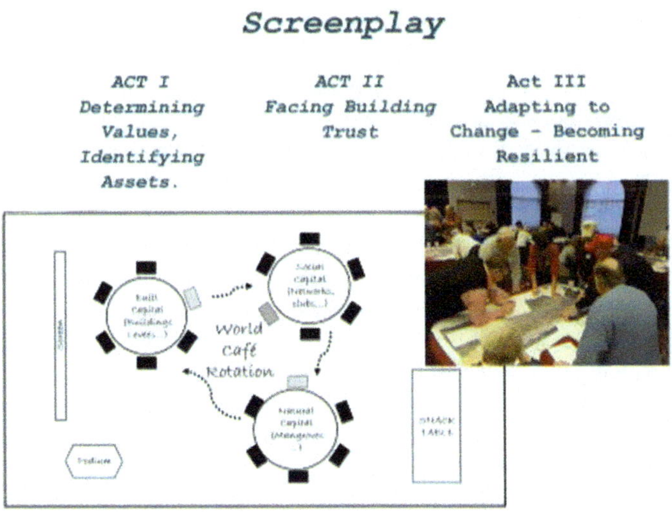

Fig. 1 The *Resilient Shores* story was preceded by a series of workshops presented as a screenplay. Each workshop exploits a world café format. Graphic prepared by, and picture taken by Bob Freitag

began with our holding a series of trust-building and direction-clarifying workshops following Hurricane Apollo.

The workshop included the use of World Cafes. These consisted of small group table discussions, as if in a cafe. However, with this process, table group members rotated, with a member staying behind to describe to the new group rotating what had transpired (Fig. 1). As a result, each group of participants added to the previous group's content. Being the timekeeper, I called for a rotation when I thought there was consensus, not agreement necessarily, but also consensus where there was discourse. The principal goal was for all to feel they had been heard, were understood, and their contributions legitimate. To trust each other. My substantive goal was to have them identify common values, or threads, if not goals.

Our public involvement continued well after Hurricane Apollo made landfall. It has never ended. We began the process with the three workshops held on City property, then it became much less formal as the discussion moved to Raj's. At Raj's Dirty Kitchen, I should say that my staff and I followed and recorded many of the deliberations more than directed them. Our current public process runs out of Belles' Community Center.

Hopefully you, by now, are familiar with the workshops that initiated our public process.

The workshops are included in the Appendix and are presented as a screenplay with three Acts representing each of the three workshops. We hope that this screenplay format will show the process used and not merely tell you the process. It may enable you to conduct workshops, perform it as an actual play, hold a table-read with the community leaders, or read the play for enjoyment. Regardless of how you approach the play, if you are involving those from your community, feel free to adapt the play to better match the perils facing you and your community.

If you conduct a table-read or hold a play, feel free to combine minor characters or split major ones among your participants. You may assign, with fewer participants, one actor to read for Eden, the condo owner, the mayor, and the council president. With larger groups, the narrator and Cassandra could be split among several community actors. Also, feel free to allow actors to tailor and expand their characters.

If you use the screenplay to guide actual community workshops, alter PowerPoint slides to better reflect your community's values, threats, and achievable aspirations.

The aim of Act I is to identify values and assets that support well-being. Your focus in Act II will be to enable participants to understand, personalize, and possibly internalize threats along with their impacts on your identified community's values. In Act III you are asking for participants to suggest alternative, achievable, acceptable pathways that reduce the threats to your identified values. He is where participants focus on their Wishes and possible Outcomes.

Most importantly, have fun. Remember that a key public involvement objective here is to strengthen networks and build trust as well as enable participants to identify values and threats and suggest a course of action. The objective here is not to develop a complete actionable plan. That takes design professionals.

Common Ground—From many stories

Conflicting Stories—apparently.

After each chapter, I told you how Paige may have told her story. Here is a summary.

> "Almost overnight I lost a place for my friends to gather," Paige began." My café was destroyed. I had thought I had lost my way of life. I was now alone. All of my Historic District business colleagues were in the same boat.
>
> "I was more fortunate to have a food truck that was stored in a safe place. It was up and running within a few days.
>
> "With no other restaurants operating and with the arrival of FEMA and State disaster recovery staff, I had all the clients I could handle. The increasing number of lookie-loos also helped.
>
> "We had been damaged before, and I knew that we would be again. I also knew that with our changing climate, things would only get worse. Many were thinking of moving away, but without other commercial interests in the District, my café would probably not survive.
>
> "The folks in the district develop two strategies with the city. Our first strategy would be to stay put and floodproof our establishment as best we could. We would form a Local Improvement District and, with the City's help, recover. Our second strategy was more long-term. IF and WHEN seas and storm surge reached the point where our businesses were no longer economically viable, we would move, as a group to a "new" Historic District in the Hilltop community.
>
> "Change would not be easy, but I would still feel I was contributing to our community, and I would generate enough income to live in Troy Breach."

Gadsden's Story could be as follows. I had one of our Tugaloo College student interns record his story from notes taken at the same basement banquet. She also completed the Resilience Story Board.

> "My independence and self-reliance were on the chopping block," may say Gadsden. "My house was damaged by Hurricane Apollo as were all the houses on the shore side of Front Street. I did not consider it destroyed as I could repair it. However, the city believed it was uninhabitable and is considering enacting a bunch of building codes. Obstacles that would make repair almost impossible. I had to move quickly and rebuild my home before the city inspectors could visit. They said that such storms were becoming more common and would get stronger. Those seas would rise, but storms have always threatened, and I have always been able to make repairs. The city established thresholds that when reached required people to move. They offered to buy the property, but I could see this coming, and they had always wanted to buy a good property.
>
> "I would continue to repair my home, and when I cannot, I will sell. They are talking about establishing a new community in Hilltop, but that does not suit me. The housing would be apartments and condos, and I cannot see myself living in condos with all the rules. No, if I cannot rebuild, I will move to some vacant land Ezra has told me about, near Valley Subdivision. Where I can pound a nail in the wall if I want to."

Developing Common Ground—A Unified, More Holistic Strategy

Troy Beach has many voices. Each faction has unique values, assets, and goals. For a community to be resilient and able to accommodate change, there needs to be an understanding of what values are commonly held and what are not. What values should be protected and defended, and which would be better minimized or forgotten?

Let's examine Paige and Gadsden's values and look for common ground. Where is the common ground? Where is it lacking, and where lacking, can acceptable strategies be found? Remember, resilience to one can be at the expense of another.

As for values, Paige values friends and social convention. Her assets include the café and her social networks. Gadsden values his independence, and he rejects what he feels are commonly held conventions in his pursuit of well-being (Fig. 2). His assets include his house and the beach. The initial recovery strategy for both is to bounce back to what was and, in the process, shore up flood reduction measures that would make recovery easier next time. Both acknowledge that there will be a next time and a next time after that. Disasters will follow disasters. Both would like help in protecting their assets. Gadsden would welcome beach nourishment and Paige a protective seawall. But that is where their agreement stops.

Gadsden questions climate change, and he would have to cross a dramatic tipping point for him to relocate. If he must leave, it would not be to the developing Hilltop community. Paige understands that the climate is changing and would not object to relocating voluntarily if remaining in place were too expensive, others in her business community were like-minded. If the Historic District would move as a group.

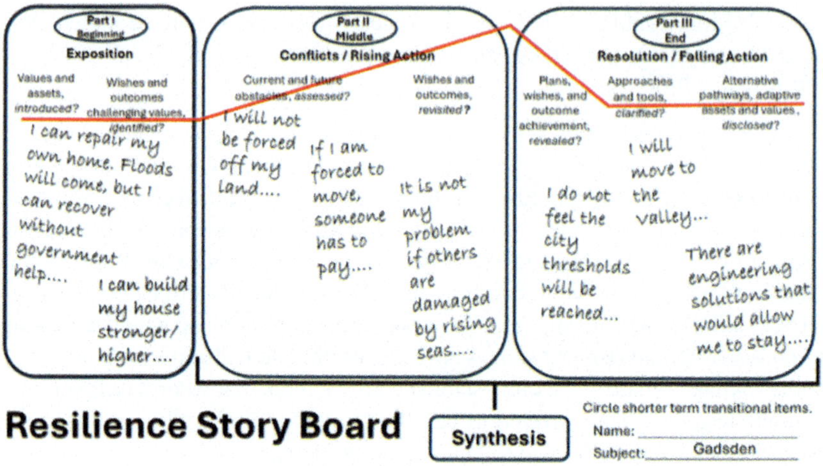

Fig. 2 Illustrates how Gadsden might have completed his Resilience Story Board. Prepared by Bob Freitag

Here are possible common values, also presented as objectives that should be agreeable to Paige and Gadsden. Strategies that do not pit one against the other.

1. Having a safe place to live—Encourage residents to reduce short-term risks to their values and safeguard their Troy Beach assets until deemed unsafe. Until robust thresholds are crossed.
2. Accepting the need for action—Agree on thresholds that, when crossed, action is necessary.
3. Having realistic choices when change is imminent—Provide a range of opportunities.
4. Striving for equity—In the case of our story, ensure that the receiving Hilltop community is not disadvantaged. Similarly, the migrating Beachers are not deprived of what they value.

These strategies address risk adaptation resulting from rising seas and increasingly frequent and intense storms. Other well-being values, such as a desire for walkability, affordable housing, shelter from heat, etc., cannot be ignored.

Story Twists

Watson's story of Troy Beach begins before the event with the landfall of Hurricane Apollo, but the story is about what occurs after the event during the Recovery phase in Emergency Management (Fig. 3). It then applied several resilience lenses, settling on a new path and implementing the Adaptive Cycle in pursuit of a new normal. The story culminated in a review of a highly managed scenario, creating a Recovery Development Corporation, implemented as extensive climate change thresholds are crossed.

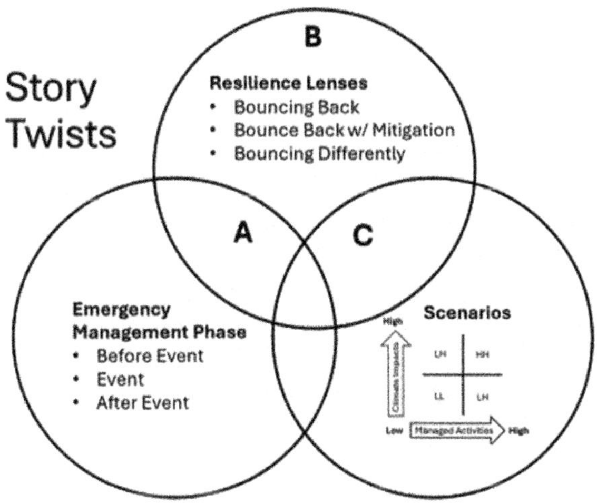

Fig. 3 Illustrated here are three sets of conditions designed to help you create your story plots. For instance, *Resilient Shores* began as a Recovery story (Circle A), Bouncing Back Differently in ultimately moving to Hilltop (Circle B), and in the process requiring considerable management skills to address future risks associated with extensive changes in our climate. Your story may take a different twist. Prepared by Bob Freitag

Gadsen's story is also one of recovery, and he follows a bounce back strategy with mitigation. He also does not accept a highly managed solution but needs to assume control.

Ozell had several stories. One followed Watson's in telling a story of increasing growth with added growth. The other possible story was in anticipation of a Troy Beach invasion. This path not taken applied a resilience lens best described as one of bouncing back with mitigation, embracing a scenario of intense management needed to protect the community from the risk of uncontrolled gentrification change.

Feel free to retell your story, with different lead characters and different twists. And remember, you can be as bold as you wish. They are just stories. However, they may reveal the best alternative for you and your community.

Considerations and Cautions

While reading Troy Beach's story, you had to make many assumptions. Extensive research was probably not undertaken, and your draft strategies should be considered hypotheses—starting points.

A key omission may be a specific risk assessment for your subject(s) of concern asset. There is extensive risk assessment guidance available. How to perform detailed risk assessments is beyond the role of this guidance. FEMA and the American Planning Association have many manuals on the subject.

My concern is that your story(s) acknowledge risk and reflect realistic future hazard perils and impacts. Hazards change over time, as do individuals and communities of individuals. This is especially true for hazards tied to climate. Most notable changes include the probable shoreward migration of storm surges (locations change), the height of the water (intensities change), and how often severe storms arrive (frequencies change).

Your subjects of concern are undoubtedly also changing. If they are cities, they may be growing and transforming. The transformation could be the result of disaster recovery, changes in population, modernization, age, or market forces. Risk determinations must reflect changes in both the hazards involved and changes of threatened assets.

As Watson and L^2 noted, future conditions are not often accounted for in Hazard Mitigation Plans. These plans often rely on FEMA flood insurance maps that are, by design, not reflective of future conditions. FEMA encourages the use of a software tool named HAZUS. Here, too, information on existing conditions is the most readily available and most often used. Those using this tool may have to add future climate or hazard information along with expected changes in the built environment.

Future land use estimates are described within Comprehensive Plans. Such was the case for Troy Beach, as Watson and Sherl noted. But as Watson relayed, at the time of Apollo, the future presented was aspirational, and climate science was not included. It was not driven by risks or risk reduction concerns. At the time of Apollo, our comprehensive plan reflected a growth strategy that encouraged high-rise development along the shore.

You should exercise caution in using existing future planning documents.

What Next?

As you read the stories of Troy Beach, Hilltop, and Valley, I offered some guidance at the end of each chapter. By this point, you may have created your stories and, in the process, suggested strategies as hypotheses. These are starting points—so what now?

Well, now may be a good time to expand your circle of influence, if you have not already done so. I am asking you to involve your community. If you cannot interest the whole community as the American Planning Association suggests, include as representative and diverse a group as possible.

You can begin your engagement following a disaster event or before the next one. You can follow this guidance as an individual or group. You can engage your family, neighbors, larger community, or city. You can work with your neighborhood organization and business colleagues or engage the Planning Department.

I hope that as your stories are told, adjusted, retold, and told again, they will offer opportunities to become real avenues for change. Troy Beach, Hilltop, and Valley may be unique communities, but their situation is not. As shoreline development owners recognize their threats, individually and collectively, they will be forced to observe alternative more difficult to accept, visions.

If your visions demand relocation, your targets for receiving areas may very likely be the more vulnerable, less expensive, service-oriented neighborhoods occupied by those who cannot afford to live in the higher-risk neighborhoods. Service workers frequently cannot afford to live in the cities where they work and must live many miles inland, often on higher ground. As such, Hilltop and Valley may be viewed as a metaphor for under-utilized communities seen as opportunities begging, according to developers, to be transformed into higher and better uses.

Hilltop may offer an alternative story—one where these safer places can take control of other stories and benefit.

I hope you enjoyed reading *Resilient Shores* and creating, telling, and retelling your stories. Thinking about how you may command change.

In parting, I want to share Watson's notes, which I found written alongside my suggestions. Some comments may be a little satirical, and some not so much. They may help you assess paths forward.

Watson's Comments on My Suggestions

1. Values drive direction.

A wealthy man built a house on the beach. He valued the view, hiked the trails, and provided a destination for his friends and family. He cut the dune to improve the view and provide sea access, directing storm surges that destroyed his house.

He relocated to the forest to enjoy the view, hike the trails, and provide a destination for his friends and family. He built a rustic cedar shake mountain cabin nestled under a beautiful mature forest canopy. He stored firewood under his porch, providing ready access to fuel a wood-burning fireplace.

What could go wrong?

Why did we move to the beach? We saw kids playing in the sand.

2. Values are often not dependent on any one single asset.

Burning a read book doesn't destroy its story.

During the COVID lockdown down I taught without a school building.

3. Common Ground

The hurricane destroyed houses, challenged livelihoods, and harmed friends. It built trust, offered new pathways, provided jobs, and strengthened relationships.

We all want children to live long and prosper, as they will for their children, and their children's children.

4. Assets can support the good, the bad, and the ugly.

Tools build houses, provide livelihoods, and craft communities. They can also destroy.

A beach community gathers support to finance and build a protective structure. They commit to the fight. The fight may also be against different, new, more reliable strategies.

5. Hazards are constantly shifting, elusive, and unrestrainable.

If you are not David, do not fight Goliath.

Rainbows have an infinite number of colors.

6. Resilience may be more art than science.

Having lots of data does not remove the need to ponder, reflect, synthesize, and dream.

7. Question Past Plans.

Plans must recognize nightmares as well as advance dreams.

Plan for what should be, not what was and can never be again.

8. Passion Rules.

Educating a slug is a waste of time.

The planner was so enlightening that she was only understood by a few colleagues.

9. Contingency Plans.

The ostrich said to himself, tradition dictates that I must stick my stupid head in the sand, but it says nothing about me NOT thinking about what must be done when I resurface.

Contingency plans represent stored energy. Opportunities.

10. Trust Governance.

Chicken Little received attention when he ran around, warning that the sky was falling. What was happening on the ground while all were looking up?

The Little Red Hen asks townspeople to gather grain for emergencies. She is ignored. All make fun of her. When there is a crisis, she:

A. ***Takes herself and her grain to a village far far way?***
B. ***Bakes bread so all smell the delicacy, sense their envy?***
C. ***Buys a gun, sits on her porch eating bread, and sneers as townspeople walk by?***
D. ***Provides her friends with bread as they come to acknowledge her foresightedness?***
E. ***Convinces all to contribute a little grain according to ability. If the event had not materialized all would have been returned. She had created a representative, diverse collection of townsfolks to establish priorities. The grain is distributed according to need?***

11. Resilience may have a darker side

A vulnerable structure is a vulnerable structure. What importance is a person's gender, color, class, or position?

Thatcher saw higher and better uses--profit--in both Valley and Hilltop. She just had to move quietly and quickly. Not that there may be something wrong with that?

Resilience often resembles a seesaw; if you go up, others may go down. And vice versa.

12. Resilience is not an end state.

Who/What does not need to be resilient? A rock? A dead guy?
Creating and sharing stories of resilience saved our troy beach communities. Could it do the same for you and your community?

Being resilient can make confronting change fun.
Before leaving, don't forget that *Resilient Shores* was written as a story, with a beginning, middle, and end. Life is not like that. Your personal stories probably won't be so exact. Tell and re-tell your stories as reflective endeavors, or to friends. Make changes, as whimsy directs, or as new facts or revisioned aspirations emerge. Pass your stories around your community, and encourage input, and corrections in the course. Create your characters to recognize that resilience has a focal point. If risks are too close to home, too threatening, you can always create an alternative safer play box, a Star Trek Enterprise or Hogwarts School of Witchcraft.

Tell your stories over dinner, drinks, and on social media. Search for strength, acceptance—common threads. Above all, have fun.

And, if you need further help, look for me. Just imagine a friendly soul wearing my bucket beach hat. Having a slice of mango pie.

I'm a good listener.

Epilogue—Well, Sort Of

Artist: Ron Kasprisin

Where's the Trojan Horse?

If you desire a comparison, the Trojan Horse could be a metaphor for our changing climate. The soldiers hiding within the horse or in this case riding atop, may be those who understand reality, but profit from exploiting the unaware. Troy, Troy Beach, is under attack.

B. Freitag, *Resilient Shores: Adapting to Climate Change Through Storytelling*, https://doi.org/10.1007/978-3-032-09047-8

Appendix A
Resilient Shores/Screenplay: Workshops

Building Trust—Table-Read/Performance/Tailoring Workshop

Right after the hurricane winds died down, the local Troy Beach press and social media were already pointing to the ineffectiveness of preparations. The local media put up a digital edition that showed houses and commercial buildings that had been damaged after each of the past storms. They included clickable before and after pictures. The houses looked better and safer after each repair. They were rebuilt to higher codes. But high tides now lapped at their footings. It was becoming more of an island home than one fronting a stable beach. The commercial buildings were the same structures damaged storm after storm.

As the community questioned their recovery, the press had latched onto the concept of resilience, and even while the storm was progressing, reporters were asking whether or not our community was able to bounce back or maybe better, bounce differently to a new future. Cassandra started wondering what resilience really meant. Was it something they should talk about for Troy Beach?

She realized resilience was too big a question for her to answer alone.

So Cassandra decided to ask the people, who are Troy Beach, how they should go forward. Not to ask for a single definitive answer. But to give them the opportunity to put their heads together, and come up with pathways that would be achievable and that they would accept. That was how the workshops came to be.

The workshops are presented as a screenplay. It is our hope that this format will better enable you to conduct your own workshops, perform it as an actual play, hold a table-readings with community, leaders, or merely read the play for enjoyment. Regardless of how you approach the play, feel free to adapt the play to better match the perils facing you and your community.

If you conduct a table-read, or play, feel free to combine minor characters or split major ones among your participants. When I performed it in class with fewer students, one student-actor read for Eden, the condo owner, the mayor, and the council

B. Freitag, *Resilient Shores: Adapting to Climate Change Through Storytelling*,
https://doi.org/10.1007/978-3-032-09047-8

president. With larger classes I assigned the roles of the narrator, and Cassandra to several students. Also, I let student tailor and expand their characters.

If you use the screenplay to guide actual community workshops, please consider replacing the PowerPoints slide information with your own to better reflect your community's values, threats, and achievable aspirations.

The screenplay has three Acts. If you have community members read the play, allow ninety-minutes to read the play. If you perform the play allow additional time, and if you conduct the play as actual community workshop, allow one day or the part of one separate day for each Act. You want each Act to spur conversations, debate, a mulling within your community.

The objective of Act I is to identify values and the assets that support them. Your emphasis in Act II will be to enable participants to understand and personalize current and future threats and their impacts on the community's values. In Act III you are asking for participants to suggest alternative, achievable, acceptable paths that reduce the threats to your identified values.

Most importantly, have fun.

Resilient Shores

ACT I—Determining Values, Identifying Assets

FADE TO:

External View. Troy Beach—Evening

A hurricane has made landfall. The larger Troy Beach community is being destroyed. HURRICANE WINDS ARE DEAFENING.

As the montage proceeds, the storm winds become softer. The camera directs attention over the shoulder of a young adult, a student intern drawing a map of the communities. The following montage is shown over the drawing, highlighting the damage in specific neighborhoods.

Note: *Participants can be shown Fig. 4 to gain a sense of the communities referenced in the screenplay.*

MONTAGE BEGINS

- Roofs of homes are torn off by hurricane winds.
- Tree branches are ripped from trees and trees have been stripped of leaves.
- Boats are shown layered across the marina.
- Storm surge has washed onto Main Street and its historic commercial structures are being destroyed.

B. Freitag, *Resilient Shores: Adapting to Climate Change Through Storytelling*, https://doi.org/10.1007/978-3-032-09047-8

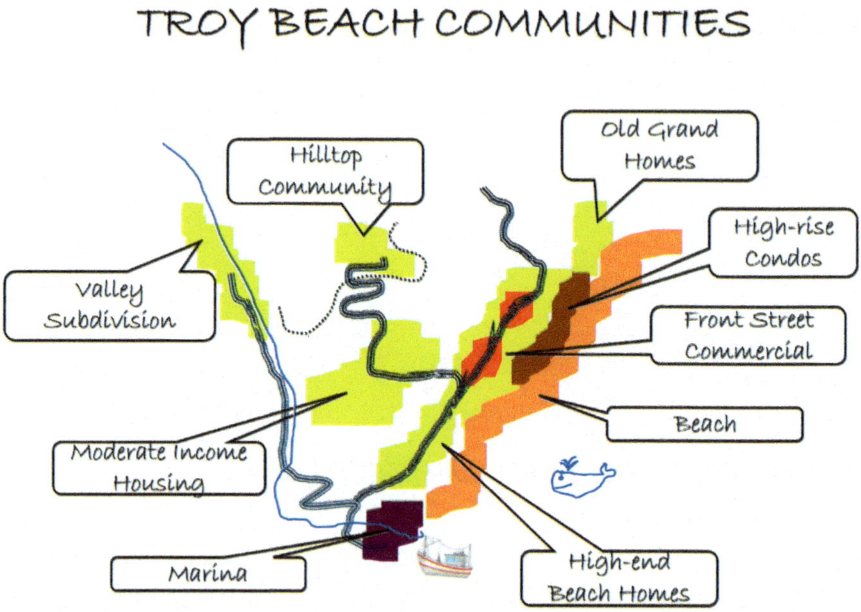

Fig. 4 The location of referenced places. Prepared by Bob Freitag

- High-end beach and older plantation homes are ripped apart.
- Patio furniture is blown off balconies.
- Waves wash onto Front Street.
- Lastly floodwaters flow into Valley Subdivision homes.

MONTAGE ENDS

OVER BLACK

TITLE: "ACT I"
  TITLE: "Threatened Values".

EXTERNAL VIEW. TROY BEACH, DAY AFTER STORM—SUNRISE

Day after the storm. ALL IS SILENT.

MONTAGE BEGINS

- Housing debris is scattered on the beach.
- Severely damaged and destroyed houses are in view.

- A child's doll is shown bouncing on the waves, washing against a storm damaged home.

MONTAGE END

EXT. OUTDOOR CANOPIES/TENTS, CITY HALL LAWN—MORNING

Tables are barely visible, partially hidden under a large tent canopy. Architectural elements are blown off the City Hall. High water marks are visible on walls. The trees in the park are void of leaves. The park lawn is brown, with white blotches of salt deposits left by evaporated seawater.

CUT TO:

INTERNAL VIEW. UNDER CANOPIES—SAME

The workshop is about to begin. Thirty-five men and women are milling about. Some are standing and others are seated at tables. Most workshop participants are dressed in casual clothes. The rest are more formally dressed attendees and include city employees, elected officials, and student interns.

Five tables with chairs are arranged neatly in front of a podium. Easels have been positioned alongside each table. A single table sign is clearly visible near each table, noting one of five well-being categories.

- Basic materials
- Health
- Good Social relations
- Security
- Freedom of choice

Each table surface is scattered with maps of the community, note pads, paper, printed Templates to document ideas and magic markers. A sixth table, also with chairs, is set off to the side. And, a seventh table, at the back of the cluster, displays an assortment of packaged snacks and drinks.

Note: *For those thinking of conducting a similar workshop, these World Cafés allow for structured conversations designed to share knowledge, offer suggestions, and build a consensus. The process offers the opportunity for individuals to work in small groups, allowing for greater depth as if discussing while in*

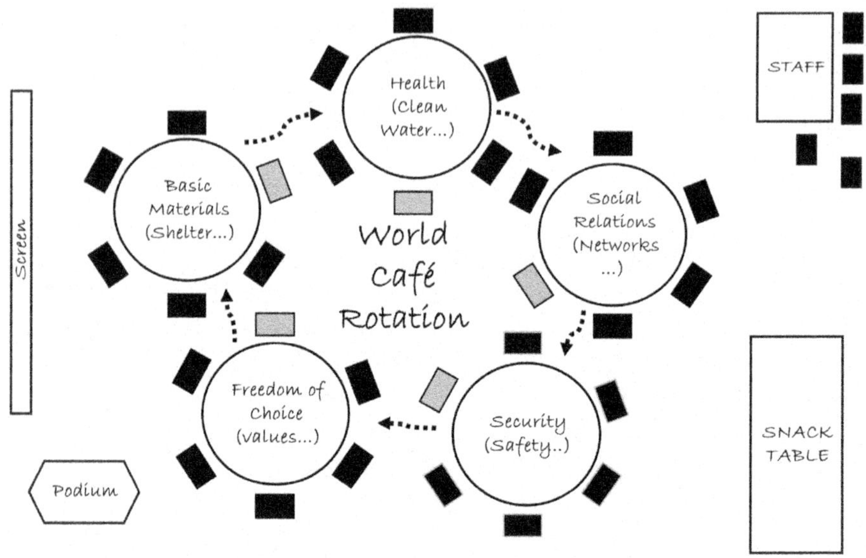

Fig. 5 Illustration of the World Café seating arrangement and room layout. Prepared by Bob Freitag

a café. Here, during this first half-day exercise, the table groups are arranged according to "well-being" categories as offered within the 2005 United Nations Millennium Ecosystem Assessment. This will also be the case for the third exercise. During the second exercise (Act II), groupings will be by "capital" categories such as built, social, and natural capital.

During each World Café exercise, table groups rotate as a group after a period. A single table lead member of the original group does not rotate and stays behind to describe previous results to a new group, thereby providing opportunities to contribute to previous work. Table groups can rotate as many times as there are table groups. Less if there are time constraints.

To make the rotating process easier, the lead could be the one rotating, although I have found that members enjoy being the ones getting up and moving.

Facilitators need to be aware of shortcomings inherent in small group discussions. Some participants may dominate while others may be passive. Some too respectful of social position, while others may be more dogmatically defiant. Workshop organizers should make themselves aware of corrective measures.

See Fig. 5 for an example of a World Café room layout.

CUT TO:

A man (CAP) walks up to a woman (CASSANDRA) who is standing at the snack table. He approaches her, somewhat aggressively.

Cassandra is the Troy Beach Planning Director and is responsible for the event. She is dressed more formally than most of the participants, but not flashy, reflecting her self-image as more of a referee of a public process than a commander. The city logo is clearly displayed on her shirt.

She doesn't have time for the approaching Cap.

CAP (STOREOWNER)

Third time's a charm, CASSANDRA. We gotta recover faster. You know, the old "Bargain Beach Wear", is my life's work, and it really got hammered this time.

CASSANDRA

(Distracted)

I know. But let's hope we can recover better this time. I really don't know if I can work in a community that seems to always set itself up for destruction.

CAP

With all your experience we trust you to get us up and running—again.

CASSANDRA

How, and how fast, we recover is up to you. Not me. And your concerns will be front and center at this workshop.

(pause)

I've always liked that shabby bucket type hat with your store's name plastered on all sides. It has, and probably still will, make ignoring you, and your concerns difficult. I am sure you'll stand out and be heard Cannot believe people still buy those.

CAP

They buy 'em. Only, I need to have an open store to sell them.

CASSANDRA

Got it. And, that's why we're here.

Cassandra leaves the snack table, and slowly walks to the podium. Cap remains.

While walking, Cassandra breaks the "FOURTH WALL".

> ***Note:*** *The term 'fourth wall' comes from the theater. If a stage is made of three walls, then the fourth wall is the imaginary, fictional one between the stage and the audience. And so, a fourth wall break is when the fiction on stage penetrates the fourth wall, out into the audience.*

CASSANDRA

(looking directly at the audience)

This guy has to realize that "we're not in Kansas anymore". If they build back quickly, they will have squandered an opportunity to create a better, more adaptive community. I'm in a real pickle if this is what the others think.

CUT TO:

INT. PODIUM—SAME

Cassandra waits at the podium for a few minutes. Without offering seating instructions attendees begin randomly sitting at tables. Cassandra visually scans those gathered and waits a short time before speaking. While waiting, she is joined at the podium by the mayor.

The mayor is wearing a suit, dressed more formally than would be expected in a beach town.

CASSANDRA

Our mayor….

MAYOR

(Interrupts)

I want to welcome you all to our MORNING workshop. This may not be Troy Beach's finest hour, but we have seen hardship before, and we are a resilient lot. You have all agreed to participate in this recovery task force and have signed a pledge to be involved for these three weeks. Thank you for your dedication. I know the sacrifices you all made. Coming here when you could be repairing your homes and businesses.

(pause)

I look forward to seeing what you come up with. You are in good hands; this is not Cassandra's first rodeo. And now that she is our Planning Director, I have great hopes.

Cap takes a seat at a table.

Cassandra returns to the podium. She thanks the mayor. She is holding her hands for comfort and is obviously nervous. Her voice quivers a little but gains strength as she proceeds.

CASSANDRA

Um... Our task this morning will be to determine what makes Troy Beach important to us. What shared values explain why we live here? What assets or things enable these values?

(pause)

And, and, how these values/assets were impacted ...changed by this most recent hurricane. Next week we will meet for the entire day and take a deeper dive into how and why OUR values are threatened.

(pause)

Um... For our third and last session, we will determine approaches ... a direction. We will do this with the understanding that we are one community. To help chart this direction we will follow a "World Cafe" type exercise format. You have each been given a table assignment.

Student interns come from the back table and distribute sheets of paper outlining table assignments. They also place a poster size sheet of paper on each table—the Troy Beach Workshop Template.

The camera zooms in, focusing on the oversized paper.

Troy Beach Workshop Template

I. Context		II. Risks	III. Adaptive Pathways		Comments
Values	Assets (Supporting Values)	Hazards / Impacts (Present / Future)	Assets (Adapted / Reorganized)	Values (Preserved, Examined, Altered)	(Champions, Capabilities, Impacts, Thresholds....)
Common Ground / Unifying Themes					

Group:________ Recorder:________

CASSANDRA (CONT'D)

Let's begin charting a new story for Troy Beach. Student interns will be at your table to help get you started and take notes. A large Workshop Template has been provided to record your notes.

A student intern facilitator will be at each table. We, staff will be available at the side table.

(Points to side-table)

We will be available to answer questions.

(pointing to an easel to the right.)

You will work in smaller groups to facilitate discussion. Today and next week, your discussion groups will be organized by Well-Being categories. Our student interns will help you to your assigned tables shortly.

(looking back to the crowd)

This morning you have two questions to discuss. They are really addressing the same question.

(Pointing to an easel, she reads the contents).

Prepared by Bob Freitag.

CASSANDRA (CONT'D)

Before we begin let's all introduce ourselves before you move to your assigned table and begin your discussion. Please use your black pens for your notes.

CUT TO:

MONTAGE BEGINS

- Participants stand and are seen introducing themselves.

- Following the introductions, participants go to assigned tables. Staff, including Cassandra go to the side table.
- Student interns go to their assigned table.
- The mayor leaves.
- Participants discuss questions, are shown taking notes and writing on easel pads.
- Table groups break and rotate between tables in a "World Café'" format when directed by Cassandra.

MONTAGE END

Actively participating in a World Cafe. Pictures taken by Bob Freitag.

CUT TO:

INT. PODIUM/SPEAKING PARTICIPANTS—LATER + 1 HOUR.

Cassandra walks to the podium from the side table. Seeing that all are actively engaged, and the process is working she smiles. The World Café has been in progress for over an hour. Groups have rotated, taken breaks, visited the snack table, and the energy level is high. After a short time she speaks. Loudly to get attention she addresses the group.

CASSANDRA

Are you all ready to share? To report to the larger group?

(pause)

Reflecting on your assigned categories, what were the values offered, and what things contributed to those values? In other words, why do you live here?

Cassandra, exhibiting greater confidence, looks at Gadsden.

Gadsden stands when noticed. A little defiantly. Chin out, leaning on his heels, words delivered slower than would be expected.

GADSDEN

Okay, I'll go first, and speak for our Health Group. There was a consensus.

(reading notes on easel)

All in our group came here because of the beach.

(Looking at Ozell)

... Well, most of us. It is where they said they get their exercise, gather shells, fishing, strolling and even seeing a whale once in a while. Where…, as many stated, they find personal peace. These are pretty much the values reflective of the group.

(pause)

… some of our group members have children and grandchildren who spend their holidays here. To share the same values. Access to the shore is really critical.

(Looking away from the easel)

I just want to stress here that we all feel that we will repair damages as we did before, quickly, before school lets out. Another major value is having a home to live in. Thanks.

While seated, and without being recognized, Ozell speaks.

OZELL (HILLTOP RESIDENT)

(In a quiet voice, but heard)

I want to make sure you all grasp that last point. We…

(pause, reflecting)

… those living in the Hilltop community, have deep roots in this area, … longer than any of you, with maybe the exception of Belle's family. We also live on the highest ground around and did okay this time. From this storm, I mean. We are fine the way things are. Right outside of the City limits.

(pause)

Just emphasizing what Gadsden said. Having a home is our greatest value.

CUT TO:

Gadsden looks puzzled. Looks at Ozell but doesn't smile.

THATCHER

(Stands without being called)

I'll go next with what our Basic Materials Group came up with. It was not much different. Although most of our discussion revolved around what drives our economy, it was still all about the beach. Because of the beach, we have an economy.

(pause)

…, And, as you plan your recovery, our group wants to make sure that you all remember that our newer buildings were not severely damaged.

PAIGE (CAFE OWNER)

(Stands when Thatcher finishes)

I've heard your stories at my cafe. Your successes and frustrations. Whatever you come up with today, you need to remember that we are a community. We can help each other. This was both our *Good Social Relations Group value* and asset. Community!

CONDO OWNER

(Standing)

We…, *our Security Group*, pretty much agrees with what was said. I have not lived here as long as many of you, and am not even a full-time resident, but my spouse and I were able to help out at the hurricane shelter and got to know many of you. What we like most, and this was said a lot during our group discussions, is that we are a community. We are really happy here, and do not want to move. We plan on retiring in Troy Beach.

CAP

(Adjusting hat for increased recognition)

Speaking for our *Freedom of Choice Group*, I just want to emphasize that the beach is why we're here in the first place. It's vital to us, and my store. We need to get our community up and running. Including getting the shore repaired ASAP…. There is a lot of variety here, lots of choices, and a lot of opportunities here. … Shopping, living, recreating, enjoying a good meal.

(pauses, notices Belle)

So, let's all work together and Let's get started.

BELLE, the community's Grande Dame, indicates she is about to speak, looking up. Most participants notice her. Room quiets down a little. Cap sits down.

BELLE

I just want to add to what Thatcher said, that for our Basic Materials Group, we also thought that our little historic district was very important to us. And if we have anything else to add it is that we are a pretty tight community with lots of shared experiences, history and culture. I do not know if I am speaking for everybody, but recovering from all these disasters seems to have made us stronger.

(pause)

…, but it kills me to see our lives disrupted so often and what we have worked for, if not destroyed, challenged every few years. I think we have to do something different.

The Volunteers Active in Disasters (VOAD) representative speaks to BELLE without being recognized. Remains seated but loud enough for all to hear.

EZRA (VOAD REPRESENTATIVE)

I have worked with most of you in my volunteer work and know you all care about Troy Beach.

(pauses, visually surveys participants)

But you also must care about your own house and that your house may be taken? Don't you deserve to keep in the family for many future generations?

BELLE

(Looking at EZRA)

Who's taking it?

EZRA

The city might not allow us to fix our homes. They might require us to move, and that would cause a lot of harm.

BELLE

First of all, who is the "they" you are referring to? Isn't the "they" actually "us?" And isn't providing a future direction why "we" are all here today? Besides, the way I see it, if we have to move, it'll be a race between local requirements, storms, waves and our changing climate.

CASSANDRA

(Assuming control of meeting)

We may be getting ahead of ourselves. This discussion, you are remembering and reflecting on, raises issues that are difficult to think and talk about. I get that.

(pause)

Over the next week please think about what was just offered? And, while you are thinking about these concerns, Eunice our climate expert, Lawrence our planning consultant, whom you all know as L-Squared, and City staff, including Gilbert our Public Works Director and me, will take what you've said today—your lists of values and contributing assets, and look at these within the context of existing and future threats.

Cassandra looks at participants, assessing attentiveness. She tightens her lips, and pauses, takes a breath visibly forcing herself not to comment on the dialogue between Ezra and BELLE.

CASSANDRA (CONT'D)

Before you go on to the rest of your day, would one of you like to try to summarize what we came up with today? To start a new story. A new vision for Troy Beach. And thank you, we'll see you all next week.

(Glances at BELLE)

BELLE nods indicating she will speak, but Allison raises her hand. Timidly. She is recognized by Cassandra and speaks softly at first, but her voice strengthens.

ALLISON

All of you are, like, missing the point. As I said within my group discussion, but wasn't reported, my value, and that of my generation, is actually having a future. Not just not having a beach to enjoy, but a livable world. You are tiptoeing around this fact. Our climate is changing and you are failing us by ignoring it. Do you disagree?

Ezra, recognizing Allison as a member of his church, raises hand, but begins talking before being recognized.

EZRA

(Condescendingly)

We are not talking about climate change today, but our community. Recovery. And as you know there are over sixty sincere and respected scientists offering alternate causes for global climate change other than deforestation and the burning of fossil fuels. But, anyway, this is not the point of these meetings.

ALLISON

(Not waiting for recognition)

They may be sincere, but, like, the vast majority of scientists agree that the world is warming and it is your fault. Me and my generation have done nothing to cause this change and it is our values that are threatened.

(Turning to Ezra)

You know I was a member of the church. I was baptized there, but lately I do not feel welcomed. Because I have been clear that I feel the church is mistaken. Denying climate change and other environmental concerns is not some form of, well, Pagan worship.

EZRA

(Again condescendingly)

This may be getting personal, we …

(turns to Allison)

Our goal here is to address Troy Beach and not solve the world's problems.

The focus returns to Cassandra. Cassandra steps away from the podium and looks at you, the reader, the audience. Us, the audience. She breaks the 'FOURTH WALL' a second time.

CASSANDRA

(talking to the audience)

Well, at least we seem to have consensus on values. It's all about the beach.

(pause)

Do you see any future problems… looming out there? With the future of the beach, with the City?

Cassandra steps back to the podium and looks at BELLE signaling that she could speak.

BELLE

(hesitates for a moment, collects thoughts)

You were in my group…

(Looks at Allison)

… and I am sorry that I didn't include your statement. Let me offer a summary statement. Okay, you want me to begin as if this morning's effort were the start of a story. Here goes.

(pause, collects thoughts, reads her notes)

Once upon a time there was a beautiful beach front community with a sandy shore, places for families to live, visitors to stay, eat and have wonderful time. Spend money. For residents to earn a decent income, raise a family and enjoy life, friends, grow old….

(looking at Allison)

And a key value underlying all of these is having a livable, if not stable, environment…. Climate.

CASSANDRA

Excellent. Thanks for this morning's work. Your values all center around the beach and having a long-term place for a home.

(pause)

We'll see you next week for our second meeting, where you will discuss the threats to what we value in Troy Beach.

CUT TO:

INT. UNDER CANOPY—SAME

Table notes and flip charts are collected. Participants stand, talk among themselves, and begin to leave the park area.

The camera shoots over an intern's shoulder, focusing on a completed workshop template.

Troy Beach Workshop Template

I. Context		II. Risks	III. Adaptive Pathways		Comments
Values	**Assets** (Supporting Values)	**Hazards / Impacts** (Present / Future)	**Assets** (Adapted / Reorganized)	**Values** (Preserved, Examined, Altered)	(Champions, Capabilities, Impacts, Thresholds....)
	Welcoming				
Community trusts	gathering places.				
each other	Open Beach, cafe				
	Community				
	centers....				
Have home for	Have a house safe				
family.	from storms.				
Family and friends	Nice Beach / shore				
want to visit.	front. Cafes and				
	shops.				
Variety of activities	Beach, Historic Dist.				
Common Ground / Unifying Themes					
Safe trusting	Beach, and welcoming				
community	businesses.				

Group: Wellbeing **Recorder:** Watson

FADE OUT:

Resilient Shores

ACT II—Facing Change, Building Trust

FADE TO:

OVER BLACK

TITLE: "ACT II"

TITLE: "Change—Weather/Climate"

EXT. DAMAGED "BARGAIN BEACH WEAR"/IMPACTED ENVIRONMENT, SAME DAY—LATE AFTERNOON

The "Bargain Beach Wear" store is shown damaged by the hurricane. Roof sections are blown off. Watermarks from flooding are clearly visible on lower remaining wall segments.

EXT. FRONT OF DAMAGED STORE—SAME

Standing by a damaged building, Cap is talking to his contractor who is dressed in working clothes.

CAP

(In familiar tone)

Just came back from a workshop that the City organized to help us recover. From the tone of the meeting, I get the feeling they may try to slow recovery. I fear that the city will pass some newfangled requirements and restrictions. I need to get this place up and running before anyone tries to stop me.

B. Freitag, *Resilient Shores: Adapting to Climate Change Through Storytelling*, https://doi.org/10.1007/978-3-032-09047-8

ATLAS (CONTRACTOR)

First of all, you are not the only fish in this pond. And, we'll need permits. It will take a little time. Just like after the last storm.

CAP

We've known each other since we were kids. You've gotta help me.

(pause)

Can you repair these exterior walls so I can take some pictures for insurance purposes? I won't claim this initial work. You can just bury these costs with later work. The codes are a bit more "flexible" with interior work, and with all of the excitement around here, who's gonna notice? Right?

ATLAS

Got your back!

CAP

I wouldn't tell the family.

CUT TO:

EXT. CITY HALL—MORNING

ONE WEEK LATER

MONTAGE BEGIN

- The City Hall architectural elements remain missing.
- High water marks are still visible.
- Workshop participants seen walking into the City Hall building. It is hotter than the previous week. Participants are visibly uncomfortable.

MONTAGE END

CUT TO:

INT. CITY HALL MEETING ROOM—SAME

Tables are arranged as before only now are located within the large City Hall conference room. A screen is in front. The podium is off to the side. The snack table has a greater assortment of packaged food items supplied by participants. On each group table are a collection of writing materials similar to the first workshop, only this time added to the collection are paper hand fans with Thatcher's business logo clearly displayed. The room is warm, and a few participants are fanning themselves displaying the corporate logo. Each table has an accompanying easel.

Cassandra walks to the podium. She looks confident. Her hands are outreaching and welcoming. She looks directly at participants. Smiles. The room is hot, and she is fanning herself using a paper fan with Thatcher's logo clearly displayed.

CASSANDRA

We will be working out of this building most of the day and we all need to thank Thatcher for the air-conditioning.

There is polite laughter.

CASSANDRA (CONT'D)

Today we will hear from Eunice

first who will discuss past and future threats to the assets and values you identified last week. What caused this most recent event and what we may face in the future? Most of you know Eunice, she is the manager of our local state park. She lives just outside of Troy Beach and I am pleased he has taken the time to speak to us.

(Pause, assessing crowd)

After Eunice's presentation and the World Café rounds, we will have lunch provided by Paige. When you all return, Gilbert, our local recovery, and floodplain manager, will discuss the disaster impacts to the assets and values you mentioned last week. We are pleased he is able to step away from his immediate recovery efforts to help with this workshop.

Eunice rises from the side table. Tables are assigned a capital- Built, Natural, and Social. She begins talking while walking to the podium. She tells participants that she will provide information then ask them to undertake a short 10 or so minute discussion. This information will be followed by requests of the participants. When each table group has discussed her last question, they will rotate in World Café fashion.

After a short pause and noticing nodding heads, she begins instructing the table groups.

Note *to those thinking of conducting a similar workshop: Realize that the information presented is illustrative of Troy Beach. When delivering your workshop,*

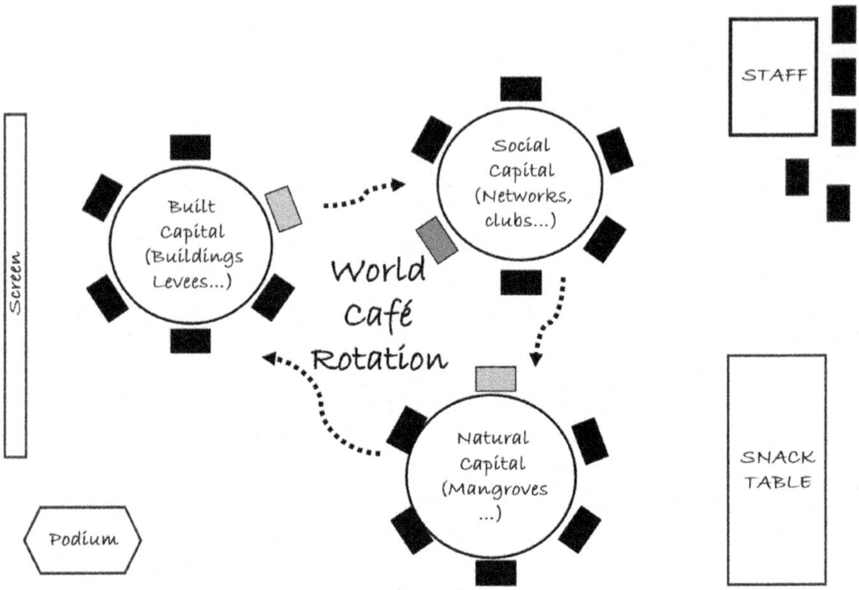

Fig. 6 Example of a World Café layout. Prepared by Bob Freitag

you will naturally provide information appropriate to your target community. Refer to Fig. 6 for a room layout possibility.

BEGIN SERIES OF SHOTS. EUNICE'S POWERPOINT PRESENTATION BEGINS. WORLD CAFÉ ROTATIONS FOLLOW

Eunice narrates her presentation.

1. Clicks: "Values/Assets"

Statement: Almost all of you said last week that you feel that our beach is the supporting asset that defines our community… The beach fulfills values such as recreation, employment, a desired lifestyle.

Request of Table group: Within your work groups, describe the relationships between the capital, the things, assets supporting the values included within your wellbeing category.

Take ten or so minutes to discuss these questions.

2. Clicks: "Values/Assets Threatened"

Statement: More damage, more often.

You have all said that we are all experiencing more damage.

Request: Threats do you fear for these categories of assets? Any local examples? This is another table group item.

3. Clicks: "Increasing Threats to Values/Assets."

Statement: Warming climate is changing our normal. Energy can change forms.

The warmer our climate gets, the more energy there is to change what we have come to expect. Our way of life. And, this energy has to go somewhere. Do something. Or be stored to something later on. You may recall this from High School physics. The law of conservation of energy.

Request: Beginning with the energy released by a hurricane, within your groups, list 3 examples of energy transformations. How it is increasing. There are interns at your table to help.

4. Clicks: "Systems, Thresholds and Tipping Points."

Statement: We all work within systems, and we all experience the impacts of crossing various thresholds within those systems. For instance:

- white ice reflects energy back into space.
- The more ice the more reflection.
- The less ice the less reflection and the warmer it gets.

A threshold is crossed when the temperature is above freezing. Ice melts. And, since ice reflects, the hotter it gets the less reflection, and with the less reflection the hotter it gets. How much flooding can you endure? Think of other systems.

Request: Suggest thresholds that may impact your lives.

5. Clicks: "Troy Beach Workshop Template" we placed on your tables so you may record your comments. Our planning team will use these materials to work from.

Request to workgroup tables: Use the questions asked to help you define your values and the assets that support these values. Suggest the threats to these assets. A template has been provided for you to record your thoughts. The student intern is available to help.

Cap stands without being recognized, ending the series. Upon seeing CAP Eunice returns to the podium.

SERIES OF SHOTS ENDS—ENDS WORLD CAFÉ ROTATIONS

CUT TO:

CAP

Eunice! I am sure that most of us here have the same question. How sure are you of these predictions?

Ezra raises hand, but before being recognized, speaks over Cap.

EZRA

…with your so called "human-induced global warming" debate…. As I said last week, there is a lot of conflicting scientific research that has to be recognized.

EUNICE

(Pausing a moment to collect thoughts)

Actually, I thought that too. But I've lived here for years like most of you. And…

(Looking at Ezra, pauses)

…how many of these storms have we survived together?

The question was meant to be rhetorical, but Ezra answers.

EZRA

As fate would have it, three, or maybe four if I count the one in the late 90's that was more a tropical depression than a hurricane. Still brought a lot of rain.

Eunice continues, looking confused, not knowing how to respond to Ezra's lack of acceptance in the science but acknowledging that events are increasing. Doesn't answer the question directly.

EUNICE

As you just said, extreme events have all been increasing. You've noticed these changes.

Eunice looks at other participants for support.

EUNICE

You all see that seas are rising, temperatures are increasing, and there are more higher frequency storms than before.

(Looks at EZRA)

This is all happening now. It is being measured now. This is not a prediction.

EZRA

I still cannot accept that it is caused by us.

EUNICE

Will you acknowledge that our "normal" is changing, and that this change will be with us for a while?

Ezra does not comment and gives a noncommittal shrug.

Cassandra, while sitting at the side table, looks at us, the audience. She breaks the 'FOURTH WALL' again.

CASSANDRA

(looking at the real or virtual camera)

That proverbial story about the blind men describing an elephant comes to mind here. Do you see ANY common ground emerging here?

Looking back at the workshop participants, Cassandra stands, speaks from the side table.

CASSANDRA

Let's all take a short break. And, when you return, you can begin adding your comments to the Troy Beach Workshop Template.

CUT TO:

INT. CITY HALL MEETING ROOM—SAME

Most participants stand and stretch. Some walk to the snack table. Voices are heard discussing the presentation.

Allison heads for the snack table. Ezra follows the student.

CUT TO:

INT. SNACK TABLE—SAME

EZRA

(Catching up to student)

I know you have problems with our church. But from our sermons I know you have heard that many in our congregation feel there is deep anti-humanism with much of the environmental movement …and the discussion around climate change. Environmentalists downplay the costs of our well-being. The costs to the poor.

ALLISON

(Softly)

With all due respect, I really disagree. And don't you see? You cannot have one without the other. The poor cannot be helped if the environment bombs. If it's, like, too hot, or there isn't water to grow food, there will be no food for the poor or anyone else. We all saw, with Covid 19, that the poor were hurt more than others.

EZRA

Let's talk later, maybe at the church. I know your family is concerned about you. We could meet at your home. Anyway, these City meetings are more about protecting Troy Beach. God's plan is God's plan.

ALLISON

(In a more confident voice)

You are missing the point. You have to admit without a healthy environment you can't have well-being. They are the same thing.

EZRA

(In a somewhat more respectful tone)

It is in God's hands.

ALLISON

(Pauses)

(Speaking more strongly)

What about Adam and the Garden of Eden, and our role as stewards…

(even stronger)

You inherited a garden but are leaving us a desert. How dare you.

The two walk back to their respective table groups. Together and still talking. Allison is walking faster as if trying to end the conversation. They each join their respective table groups.

CUT TO:

INT. VIEW. PODIUM—SAME

Gilbert moves about the podium while beginning his presentation.

GILBERT

Most of you know me. I am the Public Works Director and am responsible for keeping things running.

(pause)

Gilbert adding emphasis with animated gestures he makes the following assertions.

A few years back when the city wrote its Hazards Mitigation Plan. Troy Beach came up with many suggestions on how to reduce disaster damage. However, many of our recommendations really didn't help as much as we thought.

Our Mitigation Plan looked at existing threats to existing vulnerabilities. We then developed a way to reduce these threats as they were. Not as they would be.

The Federal Emergency Management Agency mapped our coastal 100-year floodplain. The map set a regulatory flood height we were to build to. We then recommended that you all protect your existing property to that flood height. It reflected our risk at the time the study was completed, and not what it was becoming.

Similar to Eunice's presentation, the scene fades to Gilbert giving a PowerPoint presentation. Gilbert tells the table groups that also similar to Eunice roughly 10 minutes would be allow for each discussion item he presented.

***Note** to user. When conducting your workshop, as with Eunice's presentation, you will be building your slide deck that reflects your target community.*

BEGIN SERIES OF SHOTS. GILBERT'S POWERPOINT PRESENTATION BEGINS

Gilbert, walking while talking, begins his PowerPoint presentation.

1. Clicks. "Threats"

Information provided: Troy Beach is subject to many different hazards, most of which will be made worse by our changing climate….

Request: Please jot the

threats and related impacts compromising supporting assets, hence values. Use your green colored pen to record your notes.

Now take about 10 minutes to discuss these.

2. Clicks. "Damages caused by hurricanes."

Information: our hazards damages things we have built, natural things we depend on and ways to interact with each other.

Request: Within your table groups discuss how the hurricane has impacted your social life? What capital or assets were involved? How has it impacted the natural environment? The built environment?

3. Clicks. "Flood Risks Today," Reflecting Hazards Mitigation Plan.

Information: Our Hazards Mitigation Plan reflects our risk when the Flood Hazard Map was done.
Request: Write on your template local examples where future risks may have been underestimated?

4. Clicks. "Flood Risk Tomorrow"

Information: Determining our future flood risk is difficult, but it must be done.
Request Think of what you have just experienced. Think for what Eunice said. Can you suggest likely alternative futures.
I expect you will have lively discussions, and we all look forward to your stories. The intern will help. Not to provide answers but to guide discussion.

END SERIES—WORLD CAFE ROTATIONS END

CUT TO:

Gilbert returns to the podium and offers the following summary there by ending the participation.

GILBERT (OFF CAMERA)

After all is said and done, we must conclude that our past approaches to reduce storm damage did not seem to help as much as we had thought. And as Yogi Berra from the New York Yankees, said a long time ago, the "future isn't what it used to be." The recommendations fell short. Did not incorporate future conditions.
Table group representitives present their conclusions as stories.
Gilbert remains standing and watches participant reactions. Cassandra stands while remaining at the side table acknowledging those reporting.
Workshop participants stand only as they are recognized by Cassandra. Cassandra looks at Thatcher whose hand is raised.

THATCHER

(Stands)

You may be correct. We have all noticed the damage. And what you have said is reflected with increases in my current insurance premiums. But, can't we build sewers that can withstand saltwater, and shore up our lost dunes. Aren't there things you, we, can do?

BELLE

(Stands, looks at Thatcher)

I am sure there are things we can do, but it takes money. Eunice and Gilbert, you know me, I grew up here. My mother still lives with us, and we live on the same plot of land as her mother did before her, and her mother's mother before that. We have often discussed how Troy Beach has changed. As you all have said, the beach is much smaller today. When I was a child there would be summers with no hurricane threats at all.

(Pauses)

Still, even with these threats, out-of-towners seem to be able to afford to live here. Doesn't make any sense. They come to the beach. What for… to watch it go away? We talked about this in our group last week, and it doesn't make any sense….

GADSDEN

(Interrupting)

… You say that these mitigation recommendations have not done much good. We survived. We will survive again. And sand will come back this summer as it always does. A little at least. And if seas rise, we still may have ten years or so of shore. So why not rebuild. Ten years is better than nothing and it may be longer. Even though you said you cannot be sure…

Gilbert goes closer to the podium while answering.

GILBERT

(pause)

We survived. You're right and I guess this is why so many retiring baby boomers move here. They don't seem to mind, or, don't realize what's happening. Who knows, really? Maybe in your table groups you can come up with a why.

(pause)

And to Gadsden, you may have a good idea there. Why build a structure to last a long time, if the land the building is on will be under water or won't be washed out to sea for ten or so years. Why not do what you say. Allow a use that will provide a function, and pay taxes, for say, ten or fifteen years.

(pause)

But,

(looks at Cassandra)

We may be getting off topic. We are to discuss resolutions and approaches next week.

ALLISON

(Stands after being recognized)

What you're discussing may be well and good for Troy Beach. But, like, can we forget the cause? What about discussing reducing carbon? Using solar or maybe wind energy.

CONDO OWNER

(Interrupting)

Carbon? What does that mean anyway? And it doesn't seem to affect me. I have an apartment that has no beach. My value is going down the tubes. We have to rebuild the beach.

EZRA

(Nodding)

BELLE

Maybe we can do both. If we don't try this, the changes will worsen. What we are experiencing could keep going on and on. How much warming can we take?

CUT TO:

Cassandra stands and takes the podium. Gilbert and Eunice return to the side table.

CASSANDRA

We will pick it up this afternoon. I want to thank Eunice and Gilbert. They will be available at our side table throughout the afternoon.

Cassandra pauses, looks over the assemblage.

Lunch has been provided by Paige.

MONTAGE BEGINS

- Participants stand, go to the snack table to get their lunch.
- They return to different tables.
- Participants are seen talking while eating.
- Cassandra, L2, Eunice and Gilbert walk to different tables to eat with participants,
- Student Interns move from table to table sorting supplies. As with the first day, easels are placed alongside.
- Interns place signs on each table to illustrate capital categories—"Built", "Natural", "Social" and "Economic".
- Participants walk between tables talking as they engage others.
- When finished with their meal, they deposit their waste in bins and retake their assigned seats.

- Student interns visit each table recording note written on the Poster sized Template.

MONTAGE ENDS

CUT TO:

INT. MEETING ROOM—AFTERNOON

Cassandra walks to the podium.

CASSANDRA

The questions we will explore this afternoon are how your community values such as enjoying the shore, having safe homes and shopping opportunities, and their contributing assets include your beach, houses and shops, introduced last week and discussed this morning, can be preserved. We will use the same "World Café" process. You will be working with different people during this exercise.

Cassandra writes the new table groupings on a flip chart while introducing each category.

Please go to your assigned table. Again, student interns will help you get started and take notes, but as before, will not engage in your discussion. We'll check back in an hour or so.

Table group activities are shown. People move between tables in "World Café" rounds, taking notes, discussing, writing on easels…

CUT TO:

SNACK TABLE—SAME

Thatcher, Gadsden, and Condo Owner leave their respective table groups for a break and walk to the snack table. They meet up and begin talking.

THATCHER

Okay, I get it. Our threats are increasing. Eunice is right about that. But, so is what was said earlier. There is a market here. At least for a while. And the City is still getting the tax income. Also, every time there is damage, I get paid to rebuild.

(pause)

My little secret. A win/win every time.

GADSDEN

It cost me! I do not profit from the story, like you.

(pauses, looks at Thatcher)

Not that there is anything wrong with that. We all rebuilt after the last storm. I do not get paid, but I do get some insurance payouts. We have seen changes. Storms are increasingly likely. I get it. However, there was a consensus in our group—we all want to remain in our current location as long as possible. And, as I also mentioned, maybe we can rebuild in a way that can reduce flood damage. Let me be clear, NO ONE IS GOING TO TREAD ON MY RIGHTS TO REBUILD AND STAY HERE.

THATCHER

Down boy, no one has mentioned taking your rights. At least not yet.

(GADSDEN)

(not noticing comment)

And let's not forget what Ezra said. This climate change worry may be all bunk. Things may get better.

CONDO OWNER

(Pauses, ignores bunk comment)

Unlike you, our condo did okay but the problem is that the beach pretty much disappeared. Not actually disappeared. There is a sliver left, but hardly enough in front of my condo to place a beach towel. With no beach, even with our condo being undamaged, the value may go down. Who will buy? What can we afford if or when we have to move?

(Looks at Gadsden)

And our insurance premiums are really going up. Fast.

THATCHER (DEVELOPER)

Where would you move to? Relocating to a safer site could be a good thing. I guess. But with no beach there, I see no profit. Who would want to come here? The only high ground around is the Hilltop Community. And there is no beach up there.

CONDO OWNER

(Turning to Gadsden)

Where's that?

GADSDEN

(Quietly, condescending tone)

Where the black people live. In the unincorporated County.

CONDO OWNER

(Looking away)

Oh!

Participants go back to their assigned tables while talking.

CUT TO:

CONTINUE. PODIUM—LATER, 1 + HOURS

The World Café rotations are wrapping up. There is less energy in the room. The dialogue was a little exhausting. Participants seem a little tired from their involvement. Cassandra walks to the podium.

CASSANDRA

I know it was not easy to face these questions. Please, can each of you report what you discussed?

Table group presentation results are written on flipcharts. Most speak to written text standing near the accompanying easels. Individuals seen reporting.

Cassandra steps back from table while table groups report out. She listens but says little. After about half an hour Cassandra steps in front of the podium.

CASSANDRA

We have a lot to work from. Thanks.

(Pauses)

Over this coming week I want you to mull over what you've discussed in your groups. But this time, think about what we can do to build a realistic vision of the future. How do we move forward in a way that recognizes the challenges that Gilbert and Eunice have presented—that supports the assets that preserve our values? Gilbert wants me to leave you with this thought.

(pause)

And like last time, before you go, would one of you like to build on our story?

Many in the group look at BELLE. BELLE stands without being formally recognized.

BELLE

(Pausing, reviewing her notes)

Okay I will try again.

(pause)

Once upon a time there was a beautiful beach front community enjoyed for all of the reasons stated. But it was more than a place, it was a community that had developed a strong identity capable of solving problems.

(pause)

And, then we realized, that even with its strengths, and because of a warming climate….

(pauses, looks at Allison)

… this place might not support the same values for future generations. Storms are increasing, our shore is retreating, salt water is corroding our infrastructure, our homes are being destroyed, and we are finding it more and more difficult to recover.

(pause)

But I have to add… all of this is making us a tighter, stronger community. We are learning to trust each other.

CUT TO:

Staff collect table materials. The camera focuses on the oversized Workshop Template.

Troy Beach Workshop Template

I. Context		II. Risks	III. Adaptive Pathways		Comments
Values	**Assets** (Supporting Values)	**Hazards / Impacts** (Present / Future)	**Assets** (Adapted / Reorganized)	**Values** (Preserved, Examined, Altered)	(Champions, Capabilities, Impacts, Thresholds....)
	Welcoming				
Community trusts	gathering places.	Exclusionary Practices, abandoning			
each other	Open Beach, cafe	our beach, outsiders taking control			
	Community	or hyper commercialization.			
	centers....	Families turning against each			
		other.			
Have home for	Have a house safe				
family.	from storms.	Hurricane, or storm surge destroying			
		homes and beach. Our well kept,			
Family and friends	Nice Beach / shore	welcoming Historic District.			
want to visit.	front. Cafes and				
	shops.	Storms, flooding is increasing.			
Variety of activities	Beach, Historic Dist.				
Common Ground / Unifying Themes					
Safe trusting	Beach, and welcoming	Hurricanes, risk to beach and			
community	businesses.	outsiders having control.			

Group: Wellbeing **Recorder:** Watson

FADE OUT:

Resilient Shores

ACT III—Adapting to Change, Becoming Resilient

FADE IN:

OVER BLACK

TITLE: "ACT III"

TITLE: "Becoming Resilient."

EXT. CITY HALL, WEEK LATER—MORNING.

The City Hall is visibly cleaner. Construction workers are seen on scaffolds. Participants are walking into the building in small groups. Talking.

CUT TO:

INT. MEETING ROOM—SAME.

The City Hall meeting room space is also scrubbed. The air conditioning is working and participants look more comfortable with their surroundings. There are a wider variety of homemade dishes available on snack table. Tables are arranged as they were the first day. Maps, easels, note pads printed templates to record suggestions, magic markers are visible.

INT. PODIUM—SAME.

Cassandra takes the podium. She is dressed comfortably. Fitting office informal. She looks confident. Has a big comforting and welcoming smile. The flowing announcement includes expressive hand and arm movements.

CUT TO:

B. Freitag, *Resilient Shores: Adapting to Climate Change Through Storytelling*, https://doi.org/10.1007/978-3-032-09047-8

CASSANDRA

Welcome back. This is our last workshop activity. I realize that what you discussed last week was especially difficult. This morning is a time for optimism. This morning you will address the threats and resolve the conflicts raised over the last two weeks. And, in the process create a more resilient future.

(Pause for reaction)

At the end of the workshop, you will have presented specific guidance for our professional planning team. Guidance that will be driven by the assets and values you identified during the first workshop. Guidance that recognizes existing and future threats within the context of our community's strengths and weaknesses.

(Pauses but hears no interruption)

Today you will examine possible futures. L-Squared will lead the effort. He has been hired by the city to revisit our recovery effort and comprehensive plan.

L^2 Goes to the Front and Stands by a Portable Screen. Off to the Side Are Posters Highlighting the Subject Matter and a Whiteboard. A Student Intern is Standing by the White Board Prepared to Take Notes.

L^2

As we discussed the first day, our Comprehensive Plan certainly represents our image of a residential and destination recreational community. Yet if we have learned anything from this past event and from studying our threats these past weeks, you must realize that this current vision of the future is not achievable.

L^2 Gives the Following Presentation. He Presents Information and Items to Be Discussed. Similar to Other World Café Exercises, 10 min is Allotted for the Discussion Following Each Presentation.

> ***Note*** *to User: As with other presentations, the following material would need to reflect your target community if this work is intended to guide an actual community workshop or charrette.*

BEGIN SERIES OF SHOTS. L^2S POWERPOINT PRESENTATION

L^2

Clicks: "Exiting Community Vision"—Images from the Comprehensive Plan.

Information: This is the community we have said we wanted. What most of us brought into when we moved here. A community where Condos lining the shore and services are available along Front Street….

Request: In your new groups translate your values and associated assets into wishes and goals.

You have 10 minutes.

1. Clicks: "Threats to Vision"

Information: In light of what Eunice said, it may be time for Troy Beach to re-think where it is going. Where it wanted to go. Time to reorganize.

Request Can you think of any alternative futures, trajectories? Address within your breakout groups.

2. Clicks: "Possible Future"

Information: Troy Beach has three categories of alternative futures. We can 1, re-organize and pursue a new direction. Two, we can "bounce back." Pursue our pre-event trajectory, but with better and stronger codes.

That leaves a third alternative solution, and that is to ignore the problem and take what comes. Knowing full well that what may come could be collapse.

Request: In your groups, participant offer local examples of what a future may look like for trajectories 1, 2, and 3?

3. Clicks: "Create Your Own Resolution Story"

Request: Within your table groups, review your values along with assets that enable them, in light of likely impacts to current and future threats.

Think of as many ways as possible to resolve conflicts. Ways that are acceptable to you.

You may continue using the provided Troy Beach Workshop Template to record your suggestions.

When you complete this task your story arc will be complete. You began with values and their supporting assets. And, you end with values possibly supported by modified assets.

END OF SERIES.

L^2 at the Podium. Begins His Presentation.

L^2

…In summary

One. Our Comprehensive Plan was completed after the adoption of the threat assessment. The City prepared a plan that reflected its desire to exploit our beach assets and continue our shore orientation. Building a wall of large condominiums along our shore. These recommendations are not making us resilient. These recommendations are not feasible. It is making us more vulnerable.

Two. The question now is how can we change our thinking and create a new vision?

(pause)

I will be at the side table if you have any questions. Cassandra will now begin our concluding exercise.

Questions are not requested, but Cap stands and interrupts.

CUT TO:

CAP

Sooo, I have a question. All of this "planning" will take time. Will we be allowed to rebuild or will you enact some kind of moratorium like some other communities down the road are talking about?

L^2

As long as you do not harm anybody, your rebuilding doesn't cost the City much, and you are willing to take on any future costs caused by your actions

(looks toward Cassandra)

I cannot think of any reason why you should not be able to rebuild.

(pause)

You need to discuss this in your tabled groups.

Cassandra retakes the podium.

CASSANDRA

I think now is a good time to break for lunch. I want to thank Paige for providing it.

CUT TO:

Participants are seen picking up lunch boxes from the snack table. They return to their respective tables, eating and talking in groups. There is more energy in the room.

CUT TO:

CONTINUE—AFTERNOON

Cassandra goes to the podium. When the scene begins, her introduction is in progress.

CASSANDRA

…We will now break into "World Café" well-being groups as we did our first week. Our goal here, and what we have been working toward these past weeks, is to suggest how to preserve the values that define Troy Beach for each of us.

(pause)

On your tables we have provided each group with lists of the supporting assets and values along with associated threats you prepared over the past two workshops. Now comes the creative part. The fun part.

(pause)

After you have confirmed these lists and have added anything we missed, you are to suggest as many ways as you can imagine to resolve the conflicts between threatened assets and their impact on your stated values. In some cases, you will be able to reconfigure the things/assets and support your values. And in other cases, you may have to adjust your values.

We will follow a "World Café" process as we have done previously.

(pause)

You will be sitting with different people. In different groups.

(Looking at the interns)

Can you pass out the new table assignments?

(pause)

We will now begin.

CUT TO:

Participants move to assigned tables, rotate between tables as the World Café round progresses. Take notes. City staff walks to the side table. Participants visit the table.

CUT TO:

INT. SIDE STAFF TABLE—SAME

L²

(Turning to Cassandra)

How are we doing? How'd I do?

CASSANDRA

Well, you gave the facts. Didn't recommend too many actions. Mostly just suggestions. Now it's up to them. They may want to just rebuild… Again. But I hope not. I am tired of this cycle: Damage, rebuild. Damage, rebuild…

L^2

You may have to do a little more directing. Less Socrates, and a little more Alexander the Great.

CASSANDRA

I get it.

L^2

If you use your charm, you could lead this group anywhere.

CASSANDRA

I'm not that charming. Besides, this is their community, not ours. To get a plan that works, they have to buy into it and the quickest way to do that is have them build it. We're giving them information and the way to make a blueprint. Let's see what they come up with.

Paige walks up to L^2 and Cassandra, and interrupts without hearing the previous conversation.

PAIGE

(Walking to table)

L-Squared, what are the chances of us rebuilding the dune to protect our Historic District. Several cities along the coast are building what you called "nature based" solutions to provide protection.

L^2

Well Calypso Beach spent about two million dollars several years ago for a beach renourishment project. You can see what happened. It was pretty much washed away. Not saying it didn't provide some protection, but it is no more.

(pause)

We can talk to our Senators and see if they can get us the funds. However, we are not the only pebble on the beach. This may be more difficult this time. And there would be a match required.

PAIGE

I guess it's up to us then. Could be worthwhile. Thanks.

Paige rushes back to her table group. Ozell approaches the side table. Politely waits for the conversation to end before speaking.

OZELL

(Looking at Cassandra)

What about the marina and the Vietnamese community? They have no representation here.

CASSANDRA

We tried reaching out, but those we talked to didn't see this effort being worth the time.

OZELL

I guess we have some trust issues there too. I can talk to some of my fisherman neighbors.

(pause)

They really have very few options. Have to be right at the water's edge. Maybe you could do one of these exercises with them.

(pause)

Cannot think of anything that would protect them though. Boats, nets, refrigerators.... Anyway, just an idea.

CASSANDRA

Could work if you're willing to help. Thanks.

CUT TO:

INT. WORLD CAFÉ BREAKOUT TABLE, "BASIC MATERIALS"–SAME

As typical of World Café practice, Paige was assigned to be the member of the "Basic Materials" group staying behind while the other members of her group rotated. The "Freedom of Choice" group has rotated to Paige's. They include Thatcher, Condo Owner and Gadsden. The scene begins with the Paige talking as she is sitting down joining the conversation.

PAIGE

... Well I got some ideas.

Our assets include the old rock dike protecting the older homes and the dune we re-constructed after the last hurricane. I know it has pretty much disappeared, but we should be able to rebuild both. Our assets. They enable our values. Being safe. Really need the protection.

(pause)

At least for a while.

THATCHER

Did L-Squared mention anything about cost?

PAIGE

Well, yes. It will cost a lot but we may be able to apply for Federal funds.

CONDO OWNER

We really need that dune replaced.

GADSDEN

We pay taxes. Don't we? By not rebuilding the dune, they are forcing us out of here. Well, I'm not going. And then they increase my insurance.

PAIGE

(looking at Gadsden)

The problem is that we are not the only carrot in the stew. I can think of five county beach communities in the same boat.

GADSDEN

(The looking at Eden, the Condo Owner)

We really need this project. They owe us. But, it will have to be high enough to take us out of the flood zone. Anyway, in just a couple of years I will own my home and will not be forced to buy flood insurance. They will not control me then.

CONDO OWNER

You sure? What about climate change? You sure you will be in control. Our insurance has gone up too. But for us, it is more important that we have a beach… for our grandkids while they are young. The next ten or so years are really important to us.

PAIGE

(looking at the student intern)

Let's underline what you have written—the need for both a dike and dune.

The student intern writes on the easel notepad. The table group begins discussion of other topics.

CUT TO:

OZELL'S GROUP TABLE—SAME

The Thatcher walks up to Ozell who is sitting at his group table and introduces himself.

THATCHER

(Looking down at Ozell)

We should talk. There is a possibility of you making some real dollars here. Hilltop land could become profitable.

Ozell stands while answering. He doesn't react to the demeaning introduction by Thatcher. Together they walk to the snack table. The Thatcher holds Ozell's elbow.

OZELL

Ok, let's talk.

FOLLOW TO:

SNACK TABLE—SAME

Ozell pours himself a cup of coffee while listening to Thatcher. Then speaks.

OZELL (CONT'D)

You know… Our Hilltop ancestors all moved to Hilltop after the Civil War. President Grant gave several freed enslaved families a land allotment. This program ended quickly, but no one wanted this land and we stayed. And here we live, along with some Vietnamese that relocated here in the '70s. And, our recent residents are from various places in South America. Then there's Raj who bought the old motel just off the interstate. I made an offer, but Raj got it. Probably turned out for the best.

(Pauses, looks at Thatcher)

Our houses remain the poorest, or as you may say, now, the most undervalued in the County. No one with money wanted to move here, and actually we like it that way. We're not on the beach, and that is okay too.

(looks up slightly)

The Valley Subdivision was built after the War. It's not near the beach and they solved their problem by building a swimming pool.

(pauses as of talking to self)

We folks from Hilltop could never use it, but after a few court decisions Troy Beach had to open it up to "coloreds". Hell, they closed it and filled in the pool with sand, right after. It seems that not swimming with us was more important than having a pool. Ironic, now our little neighborhood might become more valuable.

(pause)

To others.

Thatcher looks agitated. She seems about to peak. But doesn't and remains seated.

Cap walks up to the two not hearing what had transpired.

CAP

Here,

(Calling Thatcher by name)

You guys need hats.

Cap passes out hats.

OZELL

(Looking at Thatcher,
while taking a hat)

Okay, let's talk. Work this out. I think I should invite my son and the General's daughter. Nguyen. She was a childhood friend.

CAP

(Interrupting)

Good kids. They worked at the store a couple of summers. Actually, I think they were a "thing". Thought they would get married.

OZELL

Didn't work out. My son said their relationship was weird. Like he was dating his sister. Probably better as it turned out. They are still best friends. Went to undergraduate school together though. Then they separated. My son got a law degree and she got a PhD in some planning program. Still have those ridiculous hats by the way. They seem to come out whenever they get together. I expect their kids will be wearing them….

(pause)

Actually, they're both coming home next week to check on us. Seems to think we need them to help us recover from the storm. I'll invite them to our meeting.

Oh, Carlos Lopes should be invited. He lives here too. Has his own construction firm. His son got an engineering degree at State and works with some big development firm. Can't think of the name.

CAP

(With excitement)

I have got the perfect place. Also, the only place still serving full meals after the storm. Raj's motel restaurant, The Chang'en Cajun.

OZELL

Great Place. My sister is helping Raj's wife create some interesting dishes.

CAP

(Interrupts)

Ooooh… Samosas with Okra. Deep fried southern style.

BELLE walks up to the group.

CAP

(Looking at BELLE)

We're thinking about meeting at the Chang'en Cajun next week to go over a few ideas we've started talking about at this workshop. Want to join us?

BELLE

Can't wait. You'll have the opportunity to try my mango pie. I helped create it.

(pause)

Raj's wife, has an open kitchen out back for those in the neighborhood that cannot pay, or because of the storm, have lost the use of their kitchen. We often ask these guests to taste our creations, figuring that what doesn't kill them or cause them to laugh themselves to death, is a success. Some actually make it onto the menu.

CAP

(Looking at BELLE)

I'll call and make reservations. Since the storm they have been really busy.

Thatcher has been observing this discussion. When she decides to re-join the conversation, her tone exhibits new found respect.

THATCHER

The place, the menu, seems, well…, a bit sacrilegious on so many levels. At least they're not serving curried grits…. Or are they?

(pause)

Okay, I'm in. Attending the lunch that is. I'm not the best cook.

Ozell answers with a stronger, more commanding voice.

OZELL

(smiling at BELLE)

You forgot to mention when Mrs. Oh added some Kimchee to my son's Reuben sandwich. That made the menu.

(looking at Thatcher)

Well, we may not have a complete kitchen yet, but we have the start of a solid team.

Cap gives BELLE a hat and the scene closes with all putting Cap logo infused hats.

CUT TO:

INT. CITY HALL MEETING ROOM—SAME

LATER

Cassandra stands at the side table where she was seated. She walks to the podium. Participants are enjoying their involvement. There is excitement in the room. The background noise level is higher. Laughter is heard. Cassandra recognizes each table group lead.

CASSANDRA

(Speaking from the side table)

Who would like to go first?

Gadsden raises his hand. She remains standing for a while then sits down while participants are talking.

GADSDEN

I'll go first. Okay I see what you're saying. Our beach may continue to erode away and ultimately move behind the row of condos. But we should be allowed to stay…, until we cannot. A little beach renourishment, like what was done before, would help a lot. Rebuilding the rock dike would help too. Could buy us time to plan. And, if Ezra is correct, it may even end up being a permanent solution.

ROVER (MOBILE HOME OWNER)

(Interrupting)

I have to jump in here. I know I have not said anything, but I need to tell you that I had a nice mobile home in the Main Street Mobile Home Park. Got a lot of damage from the past storm. I just bought a new older home and I am going upstate to get it next week, and I should have it all connected the week after.

ALLISON

(looking at Rover)

Why didn't you move it when you heard that the storm was going to hit us?

ROVER

No time. We had only a couple of days to move it when we were told that there was a good chance Troy Beach was the target.

ALLISON

You still could have…, like, moved it.

ROVER

It takes a couple of days to get the home ready to move. We had removed the 'tongue,' and the wheels when we put it on blocks. Also, we get a half dozen or so warnings a year. Do you expect us to move each time there is a threat?

ALLISON

What about your neighbors?

ROVER

Most stayed and are in the same condition. Except for the few rich guys that had Class "A" rigs. They just drove away. A couple have already returned, but most are waiting until services are in.

ALLISON

Services?

ROVER

Water, Sewer connections. Stores, gas stations, and the like. The ones that just returned are pretty self-contained. Most even have self-composting toilets. Pricey though. Also, not all got away. Some got caught in traffic. Couldn't find a refuge.

GADSDEN

Well there's one solution. Being able to get up a move when threatened.

(pause)

I like that, but even if the old Troy Beach becomes one big mobile home park, we still need some permanently protected structures. Our Security Group also felt that the city should really help locate safer alternative places for us to move to. Places that are still somehow connected to the shore.

THATCHER

If we plan to develop a new safer location, we don't want to force relocation until existing locations become unsafe. This is what our Basic Materials Group suggests. And as you have said, so long as there is little or no adverse impact. When there is, we could look at the Hilltop Community to provide a safe place to accommodate growth and to relocate to when necessary.

(pause, looks at Ozell)

OZELL

(smiles)

THATCHER (CONT'D)

A new community site could be beneficial if it is safe and can handle the growth. Would help if it includes a viable commercial center. We might be able to allow some higher density next to the shore also, if this development had a structural life related in some way to the rate of erosion. No use in building a building to last fifty years if the land will be underwater or washed out to sea well before that. But twenty years may be fine. Could pencil out.

(pause)

Also, we would need to develop this new town site quickly so it is available before the next storm. And of course, Hilltop would have to accept annexation. As would Troy Beach.

EZRA

I really cannot believe we will need to move, or become one gigantic mobile home park. I still do not believe in what has been said about climate change. Dikes have worked in the past.

BELLE

(not ignoring Ezra)

Whatever we plan, we must make sure that we and our children can continue to enjoy the beach. It would be wonderful, as long as we are musing, if the beach were helped to recover as it was when I was a child.

THATCHER

(Interrupting)

Here is an idea. Maybe a new commercial center in Hilltop can be connected to the shore with an attractive corridor. A natural trail. We could have a free mini bus in the summer, a large golf cart or even a low impact monorail connecting the town center to the beach. Just a thought.

There is a pause. Cassandra stands.

CASSANDRA

(smiling)

Let's hear from the guy with the great hat.

CAP

(adjusts hat)

Thanks Cassandra.

(While speaking Cap throws hats to participants.)

CAP (CONT'D)

Our group doesn't really have much to add, except to reinforce what our BELLE said last week. If you wanna live here you've gotta be able to afford it. A new town site could provide less expensive housing.

PAIGE

Our store is part of our historic district. I can see we may have to move to a new location sometime, but our store and our District still has a lot of life to it. It is actually on some of the higher natural ground. We could protect it for a while at least. As was said earlier. Twenty years or until another location is available.

(pause)

Also, we have come to see value in having a mobile kitchen. Maybe we could build kitchens that can be moved or be dismantled?

Quietly, as if Ezra is talking to himself, heard by others, but not by Cassandra or other city staff.

EZRA

What's "The Plan"? The purpose of the chaos, the change, in allowing seas to rise?

Ezra pauses and Allison doesn't really interrupt, more like filling the void quickly.

ALLISON

And, if you are thinking of a new town, it could be green. There is a town in Kansas, Greensboro I believe, that went green after being destroyed by a tornado. Produces close to, like, zero carbon and cuts energy costs.

EZRA

(Standing while looking at Allison)

As I said, I still do not feel that we can do anything about climate change, but a new community center could provide some jobs. And lower housing costs. Help people. So why not try. Can't hurt. And if solar reduces the cost of energy… as I said can't hurt.

CASSANDRA

Again thank you very much. Now, as what has become our tradition, would one of you like to tell our story as it has evolved?

All look at BELLE. She notices the attention and begins to stand.

Ozell looks at her. She acknowledges. He stands.

OZELL

Let me offer a summary, this time. Try to build on our story, from the perspective of a Hilltop resident.

Once upon a time there was the community that is now called Hilltop. Its residents were a mile or so from the beach and they did not place the same value on the beach as all of you did, do. But, other than that they are pretty much the same. They want a safe place to call home, with friends, some income opportunities for them and their children, and a future. And we want to build equity for ourselves and to pass on to our children.

(pause)

However, they have some real challenges confronting them, building trust is one that we did not talk a lot about today.

(pause)

Getting back to the story. They have the same concern with our changing climate—the fear that Troy Beach might not be here to be enjoyed in their later years and most certainly there is a good chance that it will not be able to be enjoyed by their children and especially their children's children.

(pause, looks at Cassandra)

A lot of ideas were discussed here. Protecting the shore, raising homes even higher, allowing the beach and our coastal forest to reclaim areas. Protecting buildings by constructing dikes and even evolving into more of a mobile community.

(pause)

If none of this works, then there was talk of steering growth to Hilltop. And, annexation. As homes are destroyed moving to Hilltop. Well on behalf of Hilltop I want to thank you for all of the added attention. But we do not want it. Not as it is.

(pause)

Sorry, let me continuing with the story

(pause)

And then for this to happen a lot has to occur.

(pause)

And here is my proposal. If Hilltop is to be designated as some kind of climate-caused growth area, we need to build some trust. BELLE talked of community and there are seeds and probable/possible fertile ground, but we have to begin with trust.

(pause)

Let me conclude the story and offer my resolution. I am suggesting you, the City, its residents, consider the following for the plan you and your expensive consultant are preparing.

(pause)

First, Raj and his wife have started an open kitchen. I suggest that you all volunteer to help.

Second, Ezra has begun a Habitat for Humanity chapter, and they are helping our lower income residents rebuild. You could all get involved.

Third, I suggest that the Beauvoir's Fall Muster type Civil War reenactment celebrations be a little more, well… inclusive. Represent our collective heritage. Maybe next year's celebration could begin with digging out and re-opening that old Valley Subdivision swimming pool. The pool could be an incentive for growth. That includes all of us.

(pause)

Just some ideas.

Participants are initially silent then after a few moments spontaneous applause begins gradually then fills the room.

Cassandra goes to the podium.

CASSANDRA

"This past hurricane destroyed our home. Our City. More powerful storms are coming. Our climate is changing, but we are not giving up. Troy Beach will not die. We are resilient," said Cassandra, the City Planning Director as she pauses while the Council President walks to the podium.

The Council President stands where she is sitting at the side table and slowly walks to the podium, while the applause is continuing.

COUNCIL PRESIDENT

"Wow, these workshop results were more far reaching than I was expecting. Very exciting. I see them as the beginning of a new resilient vision for Troy Beach. Change is difficult. But thanks to all of your hard work and great ideas. We will save the City we have come to love.

"Over the next few months, our planners will take your recommendations, what you have defined as your values and vision, and develop a resilient strategy for our city. A vision that will capture your aspirations in the face of a changing world.

(pause)

"We hope to present this community vision at an 'Open House' in a couple of months. I hope to see you then," the Council President announces over the applause."

CUT TO:

Elected officials, staff and attendees stand and begin talking to others as they leave the room. Participants are smiling.

Staff are seen collecting table materials. Student interns take the last set of pictures as precaution. Participants are visibly active. Excited.

Materials are collected by Interns. Camera captures a completed Template.

Troy Beach Workshop Template

I. Context		II. Risks	III. Adaptive Pathways		Comments
Values	Assets (Supporting Values)	Hazards / Impacts (Present / Future)	Assets (Adapted / Reorganized)	Values (Preserved, Examined, Altered)	(Champions, Capabilities, Impacts, Thresholds....)
Community trusts each other	Welcoming gathering places. Open Beach, cafe Community centers....	Exclusionary Practices, abandoning our beach, outsiders taking control or hyper commercialization. Families turning against each other.	Having lots of community involvement activities.	Trusting / welcoming Troy Beach.	Trusting government is a problem.
Have home for family.	Have a house safe from storms.	Hurricane, or storm surge destroying homes and beach. Our well kept, welcoming Historic District.	Houses and businesses hurricane floodproofed. This could mean moving.	Safe place to live and protected shore and beach.	Many of those living close to the shore, have resources.
Family and friends want to visit.	Nice Beach / shore front. Cafes and shops.	Storms, flooding is increasing.	Protect beach and beach access for everybody.	We all take care of each other.	There are communities near by located on higher safer areas.
Variety of activities	Beach, Historic Dist.				
Common Ground / Unifying Themes					
Safe trusting community	Beach, and welcoming businesses.	Hurricanes, risk to beach and outsiders having control.	Attractive beach /welcoming place.	Trust. Living in attractive area.	Resources /Money is coming into area.

Group: Wellbeing Recorder: Watson

CUT TO:

EXT. DAMAGED "BARGIN BEACH WEAR" STORE—LATE AFTERNOON

Front of the damaged store. Scene is as it was before. There is no change in the extent of damage. ATLAS is wearing office informal attire.

CAP

Haven't you gotten started yet? What's going on?

ATLAS

Well, I started and well, the City seems to have hired some new permit officials to speed up recovery. This guy on loan to the city from Kansas came by and noticed your creativity. He slapped us with a "stop work order." There may be a fine. Anyway I had to remove the forms for the foundation of your addition. Glad he came when he did. Before we poured the concrete.

CAP

(Throwing hat to ground)

Oh, "obscenity", "obscenity." I guess, I did not need to expand anyway. This may not be the best future location for our store.

(pause)

I wonder if Cassandra knows about this.

ATLAS

Cassandra?

CAP

She's the City Planning Director. You know she saved *Retail*.

ATLAS

Your cat?

CAP

Ya. The cat was stranding on the roof of the store and I called the city. Cassandra was on weekend duty and answered the call. She activated some new volunteer firefighters who need the training.

(pause)

Funny thing is that Retail was so traumatized by all the guys clomping all over the roof chasing her in their heavy clothing and boots that she jumped off the roof onto our awning, and bounced to the ground.

(pause)

Safe as could be. Unharmed. Just ran and hid between my legs until all left.

(pause)

Cassandra was there. Initially she was all serious and stuff, and then she began laughing so hard that I thought she would be the one needing to be rescued. Could hardly breathe.

(pause)

I think I will make a date with her next week. Well, not a DATE, just a meeting.

(pause to collect thoughts)

At the Chang'en Cajun.

CUT TO:

INT. CASSANDRA'S OFFICE—DAY

Cassandra is sitting at her desk twirling her Gilligan-hat when Gilbert walks in. Hands folded behind her head. She is smiling.

GILBERT

A few of your workshop participants are meeting at the Chang'en Cajun to discuss some of the ideas introduced at the workshop. I was wondering if I could attend. I know you wanted this process to be their ball game, but there are some rules and some tools that should be explained. Allowing for the transferring of development rights is one and creating some kind of revolving loan with our local banker is another that you and I have discussed. And I would like to be present, but they have not asked directly.

CASSANDRA

No, don't get involved, not yet at least. They need to, as you would say, play the first couple of innings without us. Get a feel for the game. How am I doing with your metaphor?

(Pauses)

Cap has asked me to Chang'en Cajun. Probably for the same reason. I tried saying no, but he was very insistent. I'm interested in what they're doing but I think I should give them a little more time. What do you think?

GILBERT

I don't know, but he may have another agenda. Just saying.

Gilbert leaves. Door closes Cassandra leans back in her chair, and smiles. Puts on her ugly Gilligan hat. Looks at you. Breaking the 'FOURTH WALL' for the fifth and final time.

CASSANDRA

(to the cameral)

I'll be damned, they've made it to first base. I think we are getting there.

(pause)

Do you think we have enough material to develop a plan? Should we have pushed the community for more detail?

(pause)

Many feel there is an IF when considering climate change. Others believe it is only a matter of WHEN, and THEN it may be too late.

CUT TO:

EXT. BEACH—SUNRISE

25 YEARS LATER.

A sea turtle is crawling up the beach. Behind the turtle, beachgoers are playing volleyball beside a young emerging coastal forest. In the background, on higher ground, the Hilltop community becomes visible. The camera focuses on newer buildings and construction cranes peaking above the forest.

As the narration concludes, we see cameos of the key workshop participants.

CUT TO:

- Belle died of natural causes. Her daughter is a community organizer with the Belle Center, named for her mother. Their ancestral home was physically relocated to Hilltop.
- Cassandra was appointed Secretary of the Department of Homeland Security.
- Condo Owner died. His descents leveraged the condo's assets to buy into a modular near shore development. Their living unit was later relocated to Hilltop.
- Eunice retired and now volunteers at the Belle Community Center.
- Ezra works for Exxon Mobile. Most of his volunteer activities involve AmeriCorps Volunteers promoting and installing the use of Green and Blue energy products.
- L^2 remains active and advises clients. The firm's CEO has her PhD in planning and has deep roots within the Hilltop community.
- Gadsden lives somewhere in the Valley Subdivision. No one has seen him on the beach or visiting Troy Breach community for a long time.
- Gilbert died attempting to save hurricane partiers that had been trapped by rising surf.
- Allison is now the Director of Sustainability and Resilience for Troy Beach. She took a year's leave to be the interim CEO of the Southern Baptist Convention.
- Ozell is unofficial mayor of Troy Beach
- Paige's children work in owner's portable café kiosks scattered along the shore.
- Rover drives the trolley between the Hilltop commercial center and the beach.
- Thatcher became an advocate for modular construction. She teaches at the community college. Her development firm is owned by her son and that of Ozell.
- Atlas owns a modular construction company. He officiated the wedding between Cassandra and Cap.
- Student Interns have become planning directors and heads of planning firms guiding the construction of resilient low carbon communities along the coast. One intern wrote a story documenting Troy Beach.

CUT TO:

OVER BLACK

TITLE: "THE END"

TITLE: "Until the next challenge."

FADE OUT:

Appendix B
Cast/Production Roles

Dialogue Density: Approximate estimates of dialogue density are offered to help assign character roles. This allows you to assign more than one character to a single actor if needed.

- Very high density—over 100 lines of dialogue.
- High density—between 50 and 100 lines of dialogue.
- Medium density—between 25 and 50 lines of dialogue.
- Low—less than 25 lines of dialogue.

Narrator—Introduces each Act. (Dialogue density, medium).

Scene Headings and Action Settings Reader—A scene heading, also known as a slug line, is a line of text at the beginning of a scene that indicates the time and location of the scene. (INTERIOR, DAY…) Action setting lines are the text that enables the reader to visualize play. They tell what needs to be known to keep the story going. If possible, the unique reader should be assigned to read the Scene headings and action lines

Characters: (Except where specifically stated, gender and race are not determined. Director could assign local names.)

1. **Allison,** High School Student—Male/female. Local climate change activist in local high school. Baptized in the Southern Baptist Church of Troy Beach. Wants to create a better world. (Dialogue density, medium)
2. **Atlas**, Contractor—Male/female. A high school friend of Storeowner. A free spirit. A Self taught handyman. Evolved into a talented builder. (Dialogue density, low)
3. **Belle**, Longtime Resident—A 70+ year old white female, who was reasonably well off, but finds herself struggling, financially. She's the community Grande Dame. Her family dates to the Civil War and she still lives in the, now aging, on her ancestral plantation house overlooking an eroding bluff. Goal is to preserve her family and the community's history. (Dialogue density, high)

B. Freitag, *Resilient Shores: Adapting to Climate Change Through Storytelling*, https://doi.org/10.1007/978-3-032-09047-8

4. **Cap**, Storeowner—Male in his late 30 s. Has owned a store franchise since high school. Is fearful of risks. Goal is to maintain beach-oriented retail store. (Dialogue density, High)
5. **Cassandra/Cassander**—City Administrator/Planning Director at the time of workshops and Open House. Female/male in 30 s when the story begins. Appointed Planning Director because she/he her hard work, ability to listen, and not representing a threat to the council. Has great incite, but is often not believed, and sees it more productive to enable others, than assume control. Written as a female, but could easily be played by a male. (Dialogue density, very high)
6. **Dirty Kitchen Users**, Kitchen Users—Ms. Eniceo was a Filipina Teacher who contributed Adobo to the story. Ms. Oh was responsible for providing culinary accents to dishes. Ms. Ala'ilima created several varieties of Mango Pie that were added to the Chang'en Cajun menu. (Mentioned to add context, but none have any dialogue)
7. **Eden Family, Condo Owners**—Male/female. Part-time resident. Plans on retiring in the community. Their goal is to maintain equity. Children inherit property. (Dialogue density, low)
8. **Eunice**—A resident who does not work for the city. In this Play, he/she works for the State Parks Department. A self-described climate change expert, her goal is to raise awareness of global warming and secure a budget to address expected changes to the State Park. (Dialogue density, high)
9. **Ezra**, Volunteers Active in Disaster (VOAD) Representative—Female/male member of Volunteers Active in Disasters (VOAD). In this case, the VOAD representative is a lay leader of a faith-based group and a member of the Southern Baptist church of Troy Beach. (Dialogue density, high)
10. L^2—Planning Consultant—A&E firm contracted to prepare a recovery plan. (Dialogue density, medium)
11. **Gadsden**, Homeowner—Male/female. Moved to Troy Beach when retired. Spends a good amount of time preparing for visits from grandchildren. Money is an issue. Bought the home when it was relatively cheap. His goal is to continue to enjoy retirement living along the shore. (Dialogue density, medium)
12. **Gilbert,** Public Works Director—He/she has the principal responsibility for assuring that community infrastructure functions. Other duties include those related to the Stafford Act, the National Flood Insurance Program, and the Community Rating System. His/her goal is to get the work done. His/her main focus is on disaster recovery but helps with the workshop. (Dialogue density, medium)
13. **Mayor**—Male/Female. Weak mayor system. No political party affiliation. Pro-development. (Dialogue density, low)
14. **Ozell**, Hilltop Community Resident—Black man/woman. His/her extended family has lived in the Hilltop neighborhood since the Civil War. Hilltop is the unofficial "mayor" of the community. He/she is cautious, and a little cynical of those living outside of this tight-knit community. His/Her goal is to protect/

preserve the unincorporated community. The Hilltop community is located in the county, close to Troy Beach city limits. (Dialogue density, high)

15. **Paige**, Cafe Owner—Male/female. Bought an established restaurant with help from parents. Her goal is to enjoy her created community, and profit from the cafe. (Dialogue density, medium)
16. **President of City Council**—Male/female. Has greater political power than the mayor. Pro-development. (Dialogue density, low)
17. **Rover**, Mobile Home Resident—Male/Female. A permanent resident. His goal is to simply enjoy Troy Beach on a very limited income. (Dialogue density, low)
18. **Thatcher**, Developer—Male/female. Lives outside of Troy Beach but has developed many large condos along the coast. Sees Troy Beach as a financial asset. Her/his goal is to make money. (Dialogue density, high)

Possible Exercise Following a Table Read

Assigning participants to one of three groups corresponding to three communities. Each group is requested to develop alternative risk reduction solutions for their community as a single entity, and for at least one community member.

They are to focus on the values threatened. And only on assets, as they support values. Team groups are to identify those who benefit from the implementation of suggested measures and those who do not. They are to mitigate adverse impacts to the extent practicable. And lastly, they are to suggest both short-term and longer-term measures.

Communities include:

1. Valley subdivision.
2. Hilltop.
3. Troy Beach.

Their conclusions should be able to be described in 350 words or less.

Appendix C
Backstory and Acknowledgements

Finding Common Ground—"Live long and prosper," (Science Officer Spock, USS Enterprise) while seeming to live on a rapidly moving skateboard. (Watson)

Rule 13, "Pick the target, freeze it, personalize it, and polarize it," was introduced to the public by Saul Alinsky in 1971 as the culminating rule for radicals.[2] His "Rules for Radicals guided Labor Unions, the Civil Rights and antiwar organizing efforts of the 1960s, as well as the twenty-first Century politically Right "Tea Party" effort. Written to create power for the underclasses, his teachings now guide all shades of our political spectrum.[3,4] Alinsky's influence has not diminished, and that is why I feel the need to confront his legacy here.

Resilient Shores introduces you to the fictional Troy Beach and neighboring communities. The story is about their residents and friends, rescuing threatened values, adapting to a changing climate, and relocating to safe ground. At no time did they target opponents, freeze conflicts, personalize rivals, or polarize the community in their pursuit of a prosperous, long life.

Communities should not have to beat themselves up or others to pursue common ground.

My dad, Robert H. Freitag, never heard of Officer Spock, his common goal offering and greeting. And if he did, Spock was never mentioned to me. He knew and admired Saul Alinsky. However, he had concerns with his Rule 13. Alinsky was a Chicago-based union organizer turned community advocate in the 1950s and 1960s. Dad knew Alinsky personally, and he, too, thought of himself as a community organizer of sorts. But he worried about what resulted from implementing Alinsky's Thirteenth Rule. "I guess it is okay if the good guys create targets, personalize the actors, and polarize the situation," he would ask, "but what is the best way to prevent the other

[2] Alinsky Saul, Rules for Radicals, Random House, 1971.

[3] Quigley, Robert, Rules For The Radical Right? An Interview With Alinsky's Biographer, Mediaite, 2009.

[4] Phulwani, Vijay, The Poor Man's Machiavelli: Saul Alinsky and the Morality of Power, Cambridge University Press, 2016.

B. Freitag, *Resilient Shores: Adapting to Climate Change Through Storytelling*,
https://doi.org/10.1007/978-3-032-09047-8

guys from doing the same? And if you target them as God-forsaken good for nothing, what prevents them from targeting you? And then what are you left with—two angry groups that cannot talk to each other? Communities that have little hope of finding common ground. You end up with an armistice, not a friendship. Void of trust."

I do not feel my father was naïve. He understood the importance of power. Now, it could be deflected or redirected. But he also saw no need to "throw the baby out with the dishwater," as he often said. "No need to destroy future relationships. There was a long game. We build community at the YMCA, not adversaries."

I grew up in the YMCA. My father was the Executive Director of the Roseland Y in South Chicago and later the Chicago Hotel YMCA downtown. The Y I was introduced to was White, Protestant, and male. Catholics could easily feel unwelcome. I experienced a Y with no Jews and no boys of color. There was equity, of sorts. To ensure that all could afford to go swimming, and following the cultural norms of the era, we didn't wear swimming suits. Swimming suits were expensive, not affordable for all. They were also reportedly made of wool that clogged up the swimming pool pumps.

All this changed in the 1970s, and my dad found himself at the core of a maelstrom. Reverend Jesse Jackson filled the legacy left by Alinsky and was a civil rights community organizer with the Southern Christian Leadership Conference (SCLC). He and my Dad both felt that the Y should be more accepting. Jackson accelerated the push for civil equity in Chicago, and the Y was seen as an opportunity. My father found himself a target. Dad agreed with SCLC's goals and shared their values, but timing was an issue. SCLC felt that integration should have occurred yesterday, and my dad felt the Y needed time to adjust. Rev. Jackson was correct; the time was yesterday, and in all the discussions I had with my father, there was no question that they shared a common goal. With a common goal and a shared problem, there could be unity. Not to say that SCLC didn't have to make inequity recognizable and notorious.

Sign Exhibited at Seattle Downtown YMCA. (2024)

The good Reverend angered, challenged, and I feel was respected by, my father. Dad voiced dislike, tolerance, admiration, and fear of Jesse Jackson and his army of civil rights evangelists. This cluster of competing conflicting emotions was often reflected within the same day—a single hour. Dad feared that as the YMCA's restrictive discriminatory door cracked open, admitting a more diverse clientele, traditional financial support of the Y would proportionately flow away. My father feared nothing would be left to offer when the dust settled.

The Y has become welcoming and affirming. By the late 70s, Irish, Italians, Jews, Asians, Muslims, Mexicans, Catholics, Unitarians, and American descendants of those once enslaved were allowed to join any Y. The '70s also brought more structural parity for girls.[5] And, in the early 2000s, the Y formally rejected discrimination based on sexual orientation. Interestingly, there was less acceptance of nudity. We had to wear a swimsuit about the time I entered High School.

At no time was the Reverend Jesse Jackson, the civil rights movement, or the Y's Board of Directors described by my father as *targets* where the preferred outcome would benefit from polarization. Conflicts were personal, but I do not feel they were ever personalized. And, at the expense of being stereotypical, the Y did "overcome." There was no armistice; first came detente then camaraderie.

Exercise of Principle

Throughout my career, I've had the privilege of experiencing the importance of building communities based on shared values and goals. This principle drives our story of Troy Beach. Here are a few of those for whom I am highly indebted.

- The Philippine Rural Reconstruction Movement (PRRM). After college, I joined the Peace Corps and spent one summer with the PRRM, observing, more than helping, the organization to organize marketing cooperatives. I remember vividly those involved with my training, but regretfully, I cannot recall their names.
- Model Cities. Following the Peace Corps and graduate school, I worked for the University of Hawaii neighborhood-based planning center sponsored by their Federally funded Model Cities program. The center helped establish farmers' markets, supported the city's planning efforts, and exposed Hawaii resident students to the field of planning. We did not create targets. I appreciate Tom Dinell, Chris Zivolich, Stanley Chun, Fay Ala'ilima, Irene Fujimoto, Takashi Uesugi, and the many students associated with the program.
- Project Impact. After leaving Hawaii and working for a consultant firm in Australia, I joined HUD and the National Flood Insurance Administration. The program was absorbed into FEMA, and for the next 20-plus years, I supported mitigation and post-disaster recovery. Unfortunately, back then, our effort primarily

[5] The First Coast YMCA's former "Journey of Inclusion / Y-Pride" webpage (https://fcymca.org/diversity-and-inclusion/y-pride/, accessed in early 2022, now removed) summarized the YMCA's adoption of sexual-orientation nondiscrimination policies in the late 1990s. This institutional shift is documented in Gustav-Wrathall's *Take the Young Stranger by the Hand* (University of Chicago Press, 1998), especially p. 254, which describes the YMCA's abandonment of earlier condemnations of homosexuality by the late twentieth century.

involved helping finance the recovery of communities to a condition as close to their pre-disaster state as possible with a minor reduction in their overall risk. Before leaving FEMA in 1999, my last assignment was to direct the activities of the FEMA Region 10 Project Impact initiative. The program was created to build community-based risk networks and awareness. I am grateful to James Lee Witt, who created the program, and Seattle leaders Jim Mullen, Inez Pierce, Roger Faris, and many others.

- Cascadia Region Earthquake Workgroup (CREW). Leaving FEMA led me to become the Executive Director of CREW. CREW had similar objectives to Project Impact, only focusing on earthquake risks to small businesses. Thanks to Robert Zimmerman, Andre LeDuc, Bill Steele, Steve Charvat, Timothy Walsh, Craig Weaver, Joan Scofield, and others.
- Morgan Park High School Swim Team. Thank you, Coach Eric Hanauer, for taking a ragtag swimming team from a school with no history of victory to become City of Chicago champions. For introducing us to the value of wishing, conducting mental contrasting (Lane Tech), and constructing plans that began with our accepting grueling 6 a.m. practices. Without polarizing our adversaries and building confidence in a struggling student.

Each of these organizations created supportive networks and built social capital. Each achieved its objectives without creating divisions or polarizing the community. They began with a wish, mental contrasting, and a plan.

University of Washington Projects: Building Off Common Values and Threats to Those Values

In 1999, I left FEMA for the University of Washington Department of Urban Design and Planning and, with the assistance of Frank Westerlund, established the Institute for Hazards Mitigation Planning and Research (IHMPR). I taught courses in emergency management and, in the process, assessed my FEMA disaster efforts.

The Institute's effort to help Washington coastal communities assess their tsunami risk was critical to writing the "*Resilient Shores*" case study. Beginning in 2006, I actively sought to reduce coastal threats to Washington State Pacific-facing communities. That year, my floodplain management class researched reducing tsunami risks along the Washington coast. For many communities with little or no high ground, there was little that could be done to reduce risk.

Having no exit strategy provided the rationale for a subsequent graduate studio to be funded by the Washington Office of Emergency Management (EMD). The course objective was to determine if these high-risk tsunami-prone communities could be provided with a path to safety. Amanda Hertzfeld (Engstfeld), Katherine Killebrew, Christopher Scott, and Jean Gomez (Wiser) enrolled in my graduate-level design studio with the specific objective of determining if safety could be achieved through vertical as well as horizontal evacuation. Largely the brainchild of John Schelling, the SafeHaven Project was created to increase the life safety of Washington coastal residents. Schelling provided oversight as the Washington Emergency Management Division EMD project manager. Professor Ron Kasprisin was responsible for leading

Fig. 7 Project SafeHaven World Café—2012. Pictures taken by Bob Freitag

Research Assistants' graphically illustrate preferred vertical evacuation alternatives. Professor Omar El Anwar led a team to develop cost estimates.

Between 2010 and 2024, the University of Washington Institute for Hazard Mitigation Planning and Research executed a series of Project SafeHaven Grants. Figure 7 shows an example of one such meeting in Ocean Shores, Washington. The SafeHaven project included scoping meetings, World Cafes, analyses, and concluding open houses within all Washington coastal communities. The project suggested building vertical evacuation alternatives as well as more traditional horizontal routes to high ground where available. Reports were written, yet these were probably the least significant product.

The process involved hundreds of residents. These activities were undoubtedly widely discussed and mulled over at meals, parties, and recreation events. Reports were published and distributed, but more importantly, it is safe to assume that at the time, one would be hard-pressed to find a single resident unaware of the threat and the risk-reducing alternatives discussed.

As of this writing, with the help of vertical evacuation structures have been made available at the Westport Ocosta High School and by a Shoalwater Tribe structure in Tokeland. All the Washington coastal tribes at the foot of the Olympic Mountain range are relocating critical elements of their communities to high ground. Please see Appendix D for a listing of those involved with the SafeHaven project. It is safe to say that communities without safe ground access are considering vertical evacuation, and at-risk communities with high ground available are considering some form of migration.

While the Institute was active with the SafeHaven project, it also assisted non-coastal communities in finding common ground and developing value-protecting, asset-based threat reduction approaches as hazard mitigation plans. These efforts strengthen our Saving Troy Beach story. We wrote mitigation plans and helped deliver preparedness programs for the City of Everett, Washington. Special thanks go to the Everett City staff, including David DeHaan, Dara Salmon, Mary Schoenfeldt, and our Institute Research Assistants, Derrick Hiebert and Chasya Hiebert.

Following these projects, and with support from FEMA, the Institute applied a similar asset-based approach to assess conventional hazard mitigation and pre-disaster recovery planning across three different communities in Washington: Redmond, Everett, and the Makah tribe. The study identified community values, attributable assets, and threats to identified assets and suggested resolutions through workshops similar to those described in our Troy Beach Story. The project emphasized the importance of social capital in enhancing a community's adaptive capacity. The effort involved Professor Daniel Abramson (University of Washington), Andrew Winck (Makah Tribe), David DeHaan (Everett Emergency Manager), and Charles Gorman (Redmond Emergency Manager), along with our graduate and teaching assistants: Maximilian M. Dixon (RA lead), Tim Lehman, Jonathan Olds, Jesse Reynolds, and Chungho Kim.

At no time were residents, businesses, or government entities targeted, frozen, personalized, or polarized.

In 2017, the Institute prepared a wildfire mitigation plan for Plain Washington again using a process similar to the workshop described in the screenplay included in Appendix B. Much appreciation goes to our research team, including Lan Nguyen (Lead), Martha F. Pulido-Chavez, Harry Podschwit, Stephen Veith, and Ernesto Alvarado (co-PI).

Federal Emergency Management Agency Experience: Personalizing Recovery

Before taking a position at the University of Washington, I had a career with the Federal Emergency Management Agency, performing a series of roles culminating in being assigned the role of Federal Coordinating Officer. My involvement with two FEMA colleagues was of unique importance to the story of Troy Beach and its characterization. Richard Buck was the Federal Coordination Officer who guided the relocation of the Alaska Native communities of Allakaket and Altana. Each community was relocated to high ground in the dead of winter, safe from Altana River flooding. Richard Buck demonstrated to me that migration upland was possible, as was support from FEMA. I had a minor role in the effort.

One FEMA associate who helped me understand the pressures facing disaster survivors was Richard Balnicky, our FEMA Red Cross representative. He would continually relate, as far back as Hurricane Elaina in 1985, the extensive pressures placed on many residents of destroyed coastal homes to sell to developers. Their properties were often worth more with the house gone and even more, if neighboring homes had also been destroyed. Creating higher and higher densities on our most risk-prone lands seems to be the norm.

I appreciate Karl Kim and Eric Yamashita for asking me to write and deliver a series of coastal flood risk reduction workshops under the direction of the University of Hawaii's National Disaster Preparedness Training Center (NDPTC). And to the participants who provided valuable content for our Resilient Shores story. I thank Wayne Blanchard of FEMA and my co-authors, Susan Bolton, Dan Reichmuth, and Larry Larson, who enabled me to write a higher education Coastal Floodplain Management Course.

None of these efforts encouraged targeting, personalizing, or polarizing. The intent was to create community, impart knowledge, and build networks and social capital, not destroy it.

The Association of State Floodplain Managers (ASFPM)

From the early 1980s through the present, my involvement with ASFPM gave me the opportunity to ponder, practice, and pontificate. My role on the Board of Directors and various committees and task forces enabled me to view the greater field of floodplain management from a host of, often conflicting, perspectives. Thank you.

"*Resilient Shores*": Writing the Story

Before COVID-19, I wrote an essay on the meaning of resilience, drawing from my FEMA experience and what I had learned from the projects described and discussions with students. My targeted journal did not accept it; peer reviewers thought it needed significant revisions. About this time, I met with Jerry Murphy and David Carlton at an Association of State Floodplain Managers conference, and we decided that the faulted paper should be a book. Made sense at the time.

I prepared several drafts, and then Covid and the lockdown hit. Suddenly, I felt that writing an academic book was probably a waste of time. I then include the key points of my message as a story to my grandchildren during our weekly lockdown Zoom call. They hated it. It was too dark, with no superheroes and no satisfactory uplifting conclusion. Thanks to the Freitag family—Stevie (9 at the time), Emmitt (5), and Casper (3) and our pass the story game. I got the message and rewrote it as a survival story. Stevie was responsible for the positive effort, Emmitt for several superheroes, and Casper for forcing me to delineate a lot of detail. The effort was reworked, and thanks to Lory Peek, it was published as a short story in the Natural Hazards Review.

Still bored, along with students taking my Zoom University of Washington suite of classes, I felt there was a need to tell more about the public process I had been using for years, and in keeping with my feeling that showing was more interesting than telling, I wrote the approach as a screenplay. I also felt my students were not identifying with the very personal hardships facing survivors. Disaster preparedness, response, recovery, and mitigation were more than a collection of GIS maps, spreadsheet lists of police powers, or PowerPoint animations.

While still under lockdown with more than a bit of help from my daughter, Merete Hush, and my to-be son-in-law, Jerome Hush; I drafted a screenplay. I then forced my mitigation class students to suffer through a series of table readings. My goal in writing a screenplay was to show, and not tell, how communities of individuals could prevent the loss of community values from climate-related threats. How basing the approach on identifying commonly shared values and discussing resilience within the context of a World Café process could help build trust and be fun. Students gave the work good evaluations. Please refer to Appendix C for a list of students involved in these initial table-reads.

Jerry Murphy and Dave Carlton, who had also been involved in the play, advised me that my professional community would probably not be interested in a screenplay and suggested we write a book introducing the play. Why not? It all began with a failed paper. We also added instructions on how the readers could create their future visions of community resilience.

I then sent drafts to my friends at Island Press, and they were underwhelmed but, in the process, offered critical suggestions. One was that an old White male author and three of his similar sidekicks might benefit from some diversity.

I called on some friends and acquaintances for help. Jacki Monday made extensive editing recommendations. Ann Patton suggested organizational changes. Louvincey Brown pointed out the need to weave the greater importance of trust and culture into the narrative. Her suggestions formed the basis for Chap. 8, the last book chapter. Alan Levine offered tips on the advantages of Community Development Corporations. Andy Dannenberg made suggestions on the format. Thomas Ruppert helped with peer review.

I convinced Julie Clark, co-author of our book on floodplain management to join me, and together, we re-created and rewrote the story repeatedly.

Jerry Murphy and Dave Carlton remained involved, and the plot strengthened. Needing a more substantial visual component, I asked a UW colleague, Ron Kasprisin, to help.

Lastly, I could not have undertaken this effort without Wendy, my wife, who read and critiqued numerous drafts and suffered from the real and emotional isolation that accompanied my writing. Yet, at no time did she target, freeze, personalize, or polarize my effort.

Appendix D
Students Involved with *Resilient Shores* and Screenplay

The Troy Beach Screenplay was read by students enrolled in the 2021, 2022, and 2023 Hazard Mitigation graduate level classes. The 2021 course delivery resulted in many changes in dialogue. Amanda Hertzfeld was the lead instructor for the courses held in 2022 and 2023 where the play drove course content. In these deliveries, students assumed a character for the entire course. Hazards and community contexts were changed, and students were tasked to with developing mitigation approaches and tools for their assigned character. Hazards ranged from wildland fire, and flooding to earthquakes and tsunamis.

Daniel Acosta	Peter Gilbert Douglas	Julie Jefferson
Shabazz Abdulkadir	George Armstrong Esau	Stephen Johnson
Mahdi Alsharif	Michelle Ellsworth	Aaliya Jones
Isaac Anzlovar	Samuel Falinski	Mallory Jones
Joanne Armer	Claire Farrington	Alicia Kellogg
Matt Baldwin	William Feeney	Kevin Kunz
Renee Bialas	Amy Ferris	Heather Lam
Sunniva Feline Bloem	Josh Andrew Fider	Mathew Langley
Lucas Brisse	Clelie Ada Fielding	Paulyna Lara
Jimmy Brown	Maxwell Holmes	Nicole Larson
Felipe Borja Diaz	Max Fuangaromya	Brandon Lehnerz
Brooke Bradford	Alondra Garcia	Aaron Lecciones
Ann Bustamante	Trevor Gearhart	Albert Capilit Llaga
Nicole Canning	Matt Grosser	Allison Long
Cristina Cano-Calhoun	Maren Kristine Grunnet	Eldon Luo
Evan Patrick Carey	Meg Hamele	Nathaniel John Maloney
Leyla Church	Afroz Hasan	Aliah Marzolf
Callum Cleary	Rocky Helton	Leanna Marans

(continued)

B. Freitag, *Resilient Shores: Adapting to Climate Change Through Storytelling*,
https://doi.org/10.1007/978-3-032-09047-8

(continued)

Rhys Coffee	Stephanie Hernandez	Margot McDonald
Lauren Dee Corn	Charlotta Hohlbein	Jamie Medlin
Eddie De La Fuente	Gina Marcela Hortillosa	Joseph Minsky
Cyrus Deloye	Leah Huff	Kaydee Mittleider
Jace Denton	Chidozie Declan Iwu	Leily Mojarab
Catharina Depari	Leila Kaelyn Jackson	Merritt Moore
Allie Donahue	Carey Jang	Dylan Joseph Morrisette
Brandon Murphy	Tyler Quinn-Smith	Valerie Smith
Abigail Murray	Liz Ranieri	Nate Stevens
Gabriel Valentin Navarro	Kevin Fritz Remy	Romana Suravi
Sharmistha Nayak	Vincent Rogers	Laura Swett
Isabel Raisa Nerenberg	Aleksandr Romanenko	Anna Trevino
Meaghan O'Connor Lenth	Guion Rosenzweig	Darryl Joseph Vallejos
Aaron Olson	Ken Masaru Sadahiro	Alexis Wagner
George Oxborrow	Emily Adelia Saeger	Laura Waller
Greg Papp	Erin Irene Salle	Natalie Weiss
Caroline Passalacqua	Reba Lyn Salton	Novil Wijesekara
Sam Barry Perkins	Alison Scott	Chris Wilhelm
R. Shane Phillips	Tim Seed	Mario Zeppetelli
Cherie Quincieu	Paul Shema	
Bernadette Quinn	Rebecca Skjothaug	

Appendix E
Those Responsible for the SafeHaven Project

University Institute Team:
Students:
Jean Gomez (Wiser) (Lead)
Amanda Hertzfeld (Engstfeld) (Lead)
Josh Vitulli (Lead)
Meg Olson, (Student Lead)
Chrisopher Scott
David Smolker
Katherine Killebrew
Kirk Hochstatter
Timothy Lehman
Tricia DeMarco
Faculty:
Bob Freitag (PI)
Dan Abramson
Omar El Anwar
Editors:
Michael Godfried
Julie Clark
Washington Emergency Management Division:
Brynne Walker
Danté DiSabatino
Dave Nelson
David Nelson
ElyssaTappero
Jamie Mooney
John Schelling
Maximilian Dixon
Noemi La Chapelle
Technical Advisory Committee
Cale Ash
Frank Gonzalez
Julie Clark
Nathan Wood
Tamra Biasco
Tim Walsch
Tyree Wilde
Local Community Representatives
Andrew Winck
Charlene Nelso
Charles Wallace
Coni Wilson
Denise Rowlet
Gene Miles
Jamie Wisecup
Ken Lanfear
Lawrence Burtness
Lee Shipman
Lisa Hall
Paula Ackerman
Ronald Peregrin
Stephanie Fritz
Over 500 Residents participated in Coastal County Workshops

B. Freitag, *Resilient Shores: Adapting to Climate Change Through Storytelling*, https://doi.org/10.1007/978-3-032-09047-8

Appendix F
Templates

Resilient Story Board Template

We have created two templates to help create stories. Both were designed to help authors develop their stories as they are told and retold to friends and colleagues. Each format is intended to help direct action.

The Story Guide provides more of an outline to assist in mapping a plot, while the Story Board allows more space to develop the story. Both track Freytag's pyramid in guiding action and maintaining focus on the story's resolution(s). Both help the storyteller maintain their story arc, beginning with values and ending with them.

In *Resilient Shores*, the story plot was driven by HOW the community reached a resolution and not WHAT was accomplished. These tools were used once the storyteller had a wish, an adaptive resolve. What their community was to become provided the conflict. These templates helped to advance a quest to understand HOW Troy Beach got to the point where they could agree on a specific direction. To a specific future. A resolution supporting their values.

B. Freitag, *Resilient Shores: Adapting to Climate Change Through Storytelling*,
https://doi.org/10.1007/978-3-032-09047-8

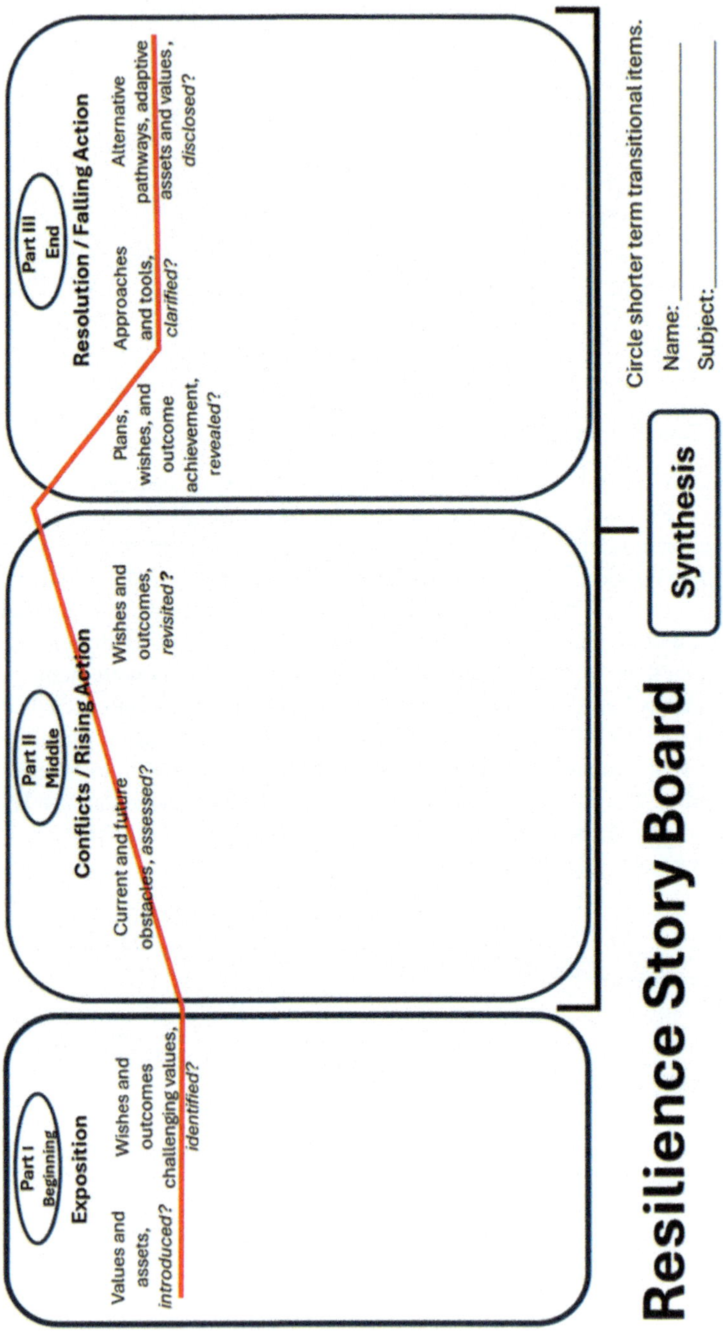
Resilience Story Board
Part I Beginning
Exposition
Values and assets, introduced?
Wishes and outcomes challenging values, identified?
Part II Middle
Conflicts / Rising Action
Current and future obstacles, assessed?
Wishes and outcomes, revisited?
Part III End
Resolution / Falling Action
Plans, wishes, and outcome achievement, revealed?
Approaches and tools, clarified?
Alternative pathways, adaptive assets and values, disclosed?
Synthesis
Circle shorter term transitional items.
Name:
Subject:

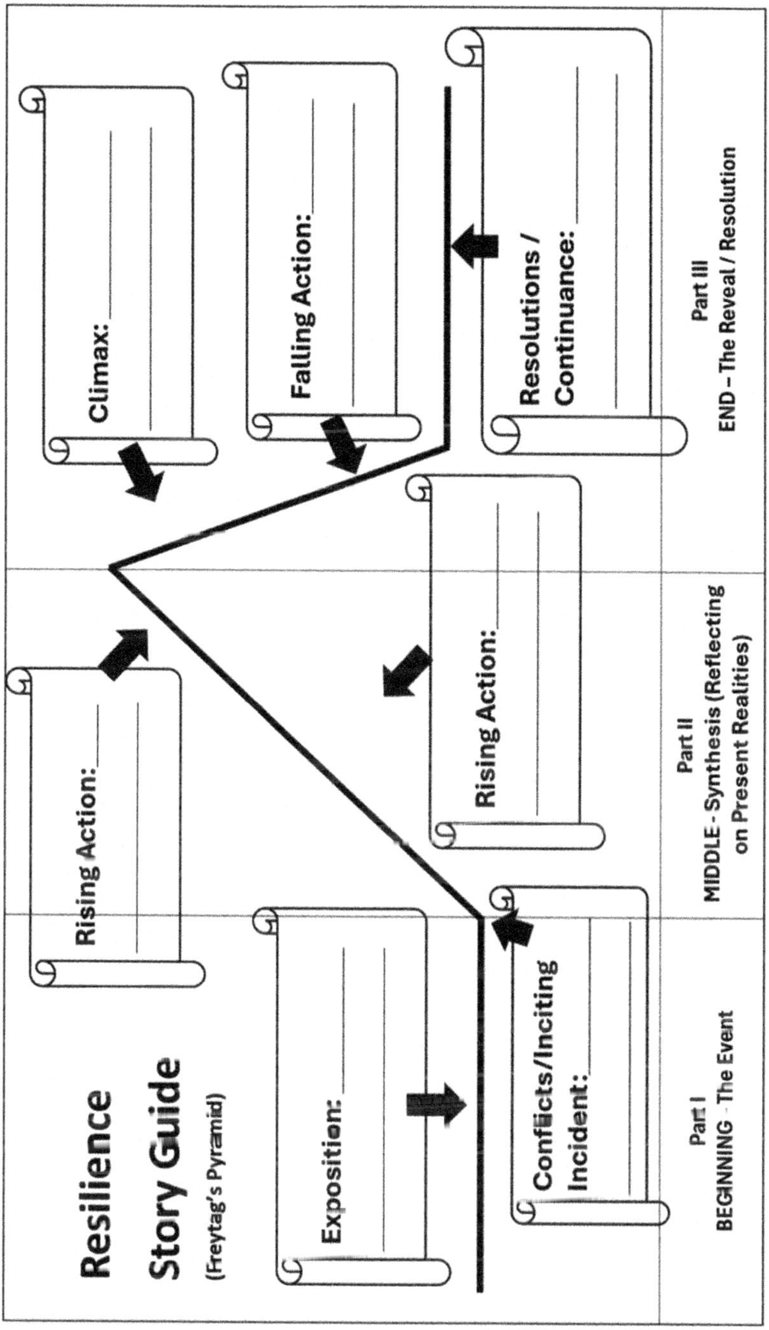
Resilience
Story Guide
(Freytag's Pyramid)
Exposition:
Conflicts/Inciting
Incident:
Rising Action:
Rising Action:
Climax:
Falling Action:
Resolutions /
Continuance:
Part I
BEGINNING - The Event
Part II
MIDDLE - Synthesis (Reflecting
on Present Realities)
Part III
END – The Reveal / Resolution

Troy Beach Workshop Recording Tool

This template was designed to help workshop participants record their thoughts while discovering WHAT strategies may provide pathways that protect their values.

The tool documents a value-driven progression. Participants begin by clarifying values and identifying the assets/capital that support them. Participants then assess threats to these asset-value relationships, suggest acceptable measures, and conclude by proposing ways assets may be adapted to better address associated values. Values typically remain consistent throughout the process, but occasionally, they may be altered in recognition of extensive threats.

In the Troy Beach screenplay, workshop participants provided their summary remarks as stories; however, the template's main purpose was to provide a record for design professionals to create an actionable plan, the content of which drove the *Resilient Shores* plot.

Troy Beach Workshop Template

I. Context		II. Risks	III. Adaptive Pathways		Comments
Values	**Assets** (Supporting Values)	**Hazards / Impacts** (Present / Future)	**Assets** (Adapted / Reorganized)	**Values** (Preserved, Examined, Altered)	(Champions, Capabilities, Impacts, Thresholds....)
Common Ground / Unifying Themes					

Group:__________ **Recorder:**_________

Appendix G
References

Abbott, H. Porter. 2002. *The Cambridge introduction to narrative.* Cambridge University Press.

Affiliated Tribes of Northwest Indians and the University of Washington Climate Impacts Group. 2024. Climate adaptation barriers and needs experienced by northwest coastal tribes: Key findings from tribal listening sessions.

Alinsky, Saul. 1971. *Rules for radicals: A practical primer for realistic radicals.*

American Institute of Planning (APA). 2020. *Climate change policy guide.*

American Institute of Planning (APA). 2022. *Equity in zoning policy guide.*

American Institute of Planning (APA). 2020. *Hazard mitigation policy guide.*

American Institute of Planning. n.d. *Local comprehensive planning framework: Interim report.*

American Planning Association. 2021. *Practice dynamic zoning.* Zoning Practice.

Bland, Lucie M., et al. 2018. Developing a standardized definition of ecosystem collapse for risk assessment. *ESA* 16(1): 1542.

Bloomfield, Emma Francis. 2024. *Science v. story: Narrative strategies for science communicators.* California Press.

Community Development Corporation. 2020. *Chinatown/international district (CID).* Seattle, WA, https://interimcda.org/

Costanza, R., et al. 2014. Changes in the global value of ecosystem services. *Science Digest* 26: 1542.

Cron, Lisa. 2023. *Wired for story: The writer's guide to using brain science to hook readers from the very first sentence.* Audiobook, Random House Audio.

Dahl, K.A., et al. 2017. Effective inundation of continental United States communities with 21st-century sea level rise. *Elemental Sciences Anthesia* 14: 546.

Dannenberg, Andrew L., et al. 2019. *Managed retreat as a strategy for climate change adaptation in small communities: Public health implications.* Springer.

Diamond, Jared. 2011. *Collapse: How societies choose to fail or succeed.* Paperback, Penguin Books.

B. Freitag, *Resilient Shores: Adapting to Climate Change Through Storytelling*, https://doi.org/10.1007/978-3-032-09047-8

Dikkers, Scott. 2014. *How to write funny*. CreateSpace Independent Publishing Platform.

Doberstein, B., J. Fitzgibbons, and C. Mitchell. 2019. Protect, accommodate, retreat or avoid. (PARA): Canadian community options for flood disaster risk reduction and flood resilience. *Nature Hazards* 98: 31–50.

Environmental Protection Agency (EPA). 2019. *Regional resilience toolkit: 5 steps to build large-scale resilience to national disasters*. EPA.

Environmental Protection Agency (EPA). *Septs to resilience video*. https://toolkit.climate.gov/#steps

Esson, Katja (Director). 2023. *Razing liberty square*. Documentary.

Fant, Charles, et al. 2021. Mere nuisance or growing threat? The physical and economic impact of high tide flooding on US roads networks. *ASCE Journal of Infrastructure Systems*.

Farber, S., and R.K. Turner. 2014. Changes in the global value of ecosystem services. *Science Digest* 26: 1452.

FEMA. *Mitigation and the emergency management cycle*. https://www.fema.gov/emergency-managers/risk-management/hazard-mitigation-planning/local-guide/introduction/mitigation-emergency-management-cycle

FEMA. 2023. *Guidance for flood risk analysis and mapping: accepting numerical models for use in the NFIP*.

FEMA. 2017. *National incident management system*.

FEMA 452. 2005. *Risk assessment: A how-to guide to mitigate potential natural disasters*.

FEMA. 2022. *State Mitigation Planning Policy Guide (3.2. Hazard Identification and Risk Assessment)*.

FEMA. 2022. *State mitigation planning key, topics bulletin: Risk, assessment*.

Fernando, Jaramillo. 2020. *Hydrological tipping points: Can we tip the bucket?* EGU BLOG.

Flyvbjerg, Bent. 2006. *Five misunderstandings about case-study research*. Sage Publications.

Freitag, Bob. 2021. Exploring a community's quest for resilience: Struggling with past plans. *Natural Hazards Review* 22(3): 545.

Freitag, Bob, et al. 2008. *Floodplain management: A new approach for a new era*. Island Press.

Freitag, B., T. Hicks, A. Jerolleman, and W. Walsh. 2020. Storytelling: Plots of resilience, learning, and discovery in emergency management. *Journal of Emergency Management* 18(5): 1542.

Freitag, Bob. 2007. How can emergency managers address our warming climate? Relying on the basics: An essay. *Journal of Emergency Management* 5(5): 12.

Godschalk, Rouch. 2015. *Sustaining places: Best practices for comprehensive plans*. American Planning Association, PAS Report 578.

Gaudiano, Brandon. 2011. Evaluating acceptance and commitment therapy: An analysis of a recent critique. *International Journal of Behavioral Consultation and Therapy*.

Georgetown Law, Georgetown Climate Center. 2020. Managed retreat toolkit. Global Environmental Change 26, 152–158. https://www.georgetownclimate.org/adaptation/toolkits/managed-retreat-toolkit/introduction.html?full

Godschalk, David R., and David C. Rouse. 2015. *Sustaining places, best practices for comprehensive planning*. American Planning Association, PAS Report 578.

Gunderson, L.H., and C.S. Holling. 2002. *Panarchy: Understanding transformations in human and natural systems*. Island Press.

Haasnoot, Marjolijn. 2020. *Dynamic adaptive policy pathways*. All ECSTATIC Materials. https://digitalcommons.usu.edu/ecstatic_all/92

Harvey, Orvis, and S. Rissman. 2018. *Designing climate solutions: A policy guide for low-carbon energy*. Island Press.

Hauer, M., J. Evans, and D. Mishra. 2016. Millions projected to be at risk from sea-level rise in the continental United States. *Nature Climatic Change* 6: 691–695. https://doi.org/10.1038/nclimate2961

Hayes, Steven, Pistorello Jacqueline, and Levin Michael. 2012. Acceptance and commitment therapy as a unified model of behavior change. *The Counseling Psychologist* 40(7): 976–1002.

Hershfield, H.E. 2011. Future self-continuity: How conceptions of the future self transform intertemporal choice. *Annals of the New York Academy of Sciences*.

Hilartner, Stephen et al. 2021. Was "science" on the ballot. *Science Magazine* 371: 893.

Housing and Urban Development (HUD), Department. 2024. *Climate resilience implementation guide: Community driven relocation*.

Huard, Jessie, et al. 2016. *Building resilience to disasters through workshops for services providers using national monitoring and evaluation framework*. Unpublished, Australian Red Cross.

Hynes, James. 2014. *Writing great fiction: Storytelling tips and techniques, course guide book*. The Great Course, The Teaching Company

NOAA. 2022. *Global and regional sea level rise scenarios for the United States*. National Oceanic and Atmospheric Administration U.S. Department of Commerce National Ocean Service, Silver Spring, Maryland.

International Panel on Climate Change Reports. 2023. *Sixth assessment report*. IPCC.

Ishiwatari, Mikio. 2022. *Managed retreat and tsunami recovery in Japan: Managed retreat: Global lessons for success*. Webinar, by NASEM Resilient America Program.

Kasprisin, Ron. 2016. *Play in creative problem-solving for planners and architects*. Routledge.

Klein, Naomi. 2008. *The shock doctrine*. Picador.

Kramer, David Kramer. 2023. Capture alone isn't sufficient to bottle up carbon dioxide. *Physics Today* 76(7): 4521.

Lerch, Daniel. 2017. *Community resilience reader: Essential resources for an era of upheaval*. Post Carbon Institute, Island Press.

Lin, Shuai Lin, et al. 2021. A method for assessing resilience of high-speed EMUs considering a network-based system topology and performance data. *Proceedings of the IMechE Part O: Journal of Risk and Reliability* 235(5): 877–895.

Lindblom, Charles E. 1959. The science of muddling through. *Public Administration Review* 19(2): 79–88.

McDaniels, T., S. Chang, D. Cole, J. Mikawoz, and H. Longstaff. 2008. Fostering resilience to extreme events within infrastructure systems: Characterizing decision contexts for mitigation and adaptation. *Global Environmental Change* 18(2): 310–318. https://doi.org/10.1016/j.gloenvcha.2008.03.001

McKee, Robert. 1997. *Story: Substance, structure, style and the principles of screenwriting hardcover*. Regan Books.

Mendenhall, Allyson, et al. 2017. *Creating planning documents.* American Planning Association, PAS Report 589.

Mileti, Dennis. 1999. *Disasters by design: A reassessment of natural hazards in the United States.* Washington, DC: Joseph Henry Press.

Millennium Ecosystem Assessment (MEA). 2005.

Mortreux, Colette. 2018. Political economy of planned relocation: A model of action and inaction in government responses. *Global Environmental Change.*

Murphy, Gerald, et al. 2023. No adverse impact legal guide for flood risk management. *ASFPM* 23(6).

National Aeronautics and Space Administration (NASA). 2022. *Rising sea level could exceed estimates for U.S. coasts—climate change: Vital signs of the planet.* NASA.

National Academy of Sciences (NAS). 2012. *Disaster resilience: A national imperative.* NAS Press.

National Academy of Sciences. 2024. *Community-driven relocation: Recommendations for the U.S. gulf coast region and beyond.* NAS Press.

National Institute of Building Sciences. 2005. *Natural hazard mitigation saves: An independent study to assess the future savings from mitigation activities.*

New Jersey, Department of Environmental Protection. 2025. *Greenhouse gas emissions.* https://dep.nj.gov/ghg/

Nigeria Conference Paper. 2015. *Building flood disaster resilience of SMEs in urban cities.*

Oettingen, Gabriele. 2025. *Positive thinking inside the new science of motivation.* Penguin Random House.

Peterson, Garry D., Graeme S. Cumming, and Stephen Carpenter. 2000. Scenario planning: A tool for conservation in an uncertain world, conservation biology. *Center for Limnology.*

Phulwani, Vijay. 2016. *The poor man's Machiavelli: Saul Alinsky and the morality of power*. Cambridge University Press.

Quigley, Robert. 2009. Rules for the radical right? An interview with Alinsky's biographer. *Mediaite.*

Rand, Ayn. 1996. *The fountainhead*. Signet, 1996.

RavenGeoffrey, Kirk, et al. 1984. *The presocratic philosophers.* Cambridge University Press.

Rojas, James, and John Kamp. 2022. *Dream play build; hands-on community engagement for enduring spaces and places*. Island Press.

Schwab, James (Editor). 2014. *Organizing for recovery planning in advance of the plan*. APA PAS 576.

Seinfeld, Jerr. 1993. *The outing*.

Sherif, Muzafer, et al. 1961. *The robbers cave experiment*. University of Oklahoma, The Institute of Group Relations.

Siders, S., et al. 2018. Managed retreat in the United States. In: *Disaster research center, biden school of public policy and administration, geography and spatial sciences*. University of Delaware, Newark. https://doi.org/10.1016/j.oneear.2019.09.008

Siders, A.R. Types of acquisition strategies. In *University of Pennsylvania, Perry world house global symposium, made possible in part by a grant from the Carnegie corporation of New York*.

Snyder, Blake. 2005. *Save the cat*. McNaughton and Bunn, Inc.

Southern Baptist Annual Meeting. 2007. *Resolution on global warming*.

Steinbeck, John. 1952. *East of eden*. Penguin, 1952 ("For if 'Thou mayest'—it is also true that 'Thou mayest not'." Chapter 24).

Stern, Marc J. 2018. Social science theory for environmental sustainability. In *Techniques in ecology and conservation*. Oxford University Press.

Stoetzer, Lasse S., and Florian Zimmermann. 2024. A representative survey experiment of motivated climate change denial. *Nature Climate Change*.

Swift, Johnathan. 1729. A modest proposal. In *Project Gutenberg 2008*.

Thaler, Thomas. 2021. Just retreat—how different countries deal with it: Examples. *Journal of Environmental Studies and Sciences* 11: 412–419. https://doi.org/10.1007/s13412-021-00694-1

Theroux, Paul. 2016. *Deep south: Four seasons on back roads*. Mariner Books.

Tuckman, Bruce W. 1963. Developmental sequence in small groups. *Psychological Bulletin* 63(6): 384–399. 10.1037/h0022100.

Union of Concerned Scientists. 2014. *Encroaching Tides: How Sea Level Rise and Tidal Flooding Threaten U.S. East and Gulf Coast Communities over the Next 30 Years*, by Erika Spanger Siegfried et al. Cambridge, MA: Union of Concerned Scientists.

Union of Concerned Scientists. 2017. *When Rising Seas Hit Home: Hard Choices Ahead for Hundreds of U.S. Coastal Communities*, by Erika Spanger Siegfried, Kristina Dahl, Astrid Caldas, and Melanie Fitzpatrick Cambridge, MA: Union of Concerned Scientists.

Union of Concerned Scientists. 2018. Underwater rising seas, chronic floods, and the implications for US coastal real estate. *UCS*.

Vobel, Kenneth P. 2010. *Right loves to hate, imitate Alinsky*. POLITICO.

Vonnegut, Kurt. 1999. *Slaughterhouse-five. ("So it goes.") Dial Press Trade Paperback*.

Walker, B., and D. Salt. 2012. *Resilience thinking: Sustaining ecosystems and people in a changing world*. Washington, DC: Island Press.

White and Epston. 2023. *Narrative means to therapeutic ends.* W. W. Norton & Company.

Wing, O.E.J., W. Lehman, P.D. Bates, et al. 2022. Inequitable patterns of US flood risk in the Anthropocene. *Nature Climate Change.*

World café. http://www.theworldcafe.com/

World Health Organization. 2024. *The power of storytelling for health impact.* WHO.

Writers.com. Freytag Pyramid.

Yodo, Nita, and Wang, Pingfeng. 2016. Engineering resilience quantification and system design implications: a literature survey. In *Design automation conference, proceedings of the 53rd annual design automation conference.*

Appendix H
Author/Contributors

Author

Bob Freitag was the director of the Institute for Hazards Mitigation Planning and Research (IHMP) and an instructor in the Department of Urban Design and Planning at the University of Washington (UW). Recognized for his contributions to education and disaster planning, he has received the Outstanding Part-time Instructor Award from the department and the Jerry Louthain Award for Distinguished Service from the Association of State Floodplain Managers. He has developed and taught courses for the Federal Emergency Management Agency (FEMA), National Disaster Preparedness Training Center (NDPTC), and the University of Washington, specializing in community resilience, hazards mitigation and floodplain management. Before joining UW, he served with FEMA, deploying on over 50 Federal Disaster Declarations. His final role with FEMA Region 10 was as a Federal Coordination Officer, overseeing response efforts.

Contributors

J.L.S. Clark is a member of Willamette Writers. (1,600 members). She used her Political Science degree in the first part of her professional life in political campaigns as well as at the Oregon Legislature. She also worked at various state and local governments. After that, she earned an MS in Geology and managed the Information Section of the Oregon Department of Geology and Mineral Industries, including overseeing the production of earthquake maps with important advances in understanding of Oregon's earthquake risk, and the magazine, "Oregon Geology." She also served as the Lead Report Preparation Consultant on the FEMA publication "Vertical Evacuation from Tsunamis: A Guide for Community Officials" (FEMA P646A), now in its second edition; and co-authored the book, "Floodplain Management: A New Approach for a New Era." Washington, DC: Island Press, 2008.

Jerry Murphy graduated from the University of Florida in 2000 with a J.D. with honors. Jerry is a certified floodplain manager, certified urban and regional planner,

B. Freitag, *Resilient Shores: Adapting to Climate Change Through Storytelling*, https://doi.org/10.1007/978-3-032-09047-8

and legal scholar. His work and research involve land use planning and control law, community-based planning, code and regulation drafting, floodplain management, growth and infrastructure finance, post-disaster redevelopment planning, community resilience/sustainability. He currently works with local governments through UF to assemble teams of expert faculty, researchers, and student assistants to craft planning approaches and strategies to address climate change, extreme weather, sea-level rise, "peril of flood" and the other challenges facing local governments in the twenty-first century.

Dave Carlton is the founder of DKCarlton and Associates, PLLC, a firm specializing in water resources, floodplain and stormwater management, and habitat restoration/preservation. He has been an adjunct instructor teaching floodplain management classes at the University of Washington. He has over 40 years of experience working with communities across the country to address their floodplain and stormwater issues. He has authored numerous plans for communities within Washington to help them address their flooding and habitat issues. He is a recipient of the Association of State Floodplain Managers' Jerry Louthain Award for Distinguished Service to Members, is a past committee chair and is past chair of the Certified Floodplain Manager Exam Workgroup for the Association. For the past twelve years he has been one of two primary consultants to FEMA on their Community Rating System, a program designed to reward communities for good floodplain management. He has taught courses on floodplain management for the CRS program across the country and continues to be involved in the improvement of the program.

Ron Kasprisin is a Signature Member of the Northwest Watercolor Society. His watercolors reflect years of work and travel throughout the Pacific Northwest, British Columbia and Alaska as an architect/urban planner working with waterfront towns and cities along the great Inside Passage from Seattle to Haines/Skagway to Fairbanks and the Yukon River pipeline bridge.

Index

B. Freitag, *Resilient Shores: Adapting to Climate Change Through Storytelling*,
https://doi.org/10.1007/978-3-032-09047-8

The manufacturer's authorised representative in the EU is Springer Nature Customer Service Centre GmbH, Europaplatz 3, 69115 Heidelberg, Germany. If you have any concerns regarding our products, please contact ProductSafety@springernature.com

Printed and bound by CPI Group (UK) Ltd, Croydon, CR0 4YY
17/09/2026
02218437-0004